Biomineralization

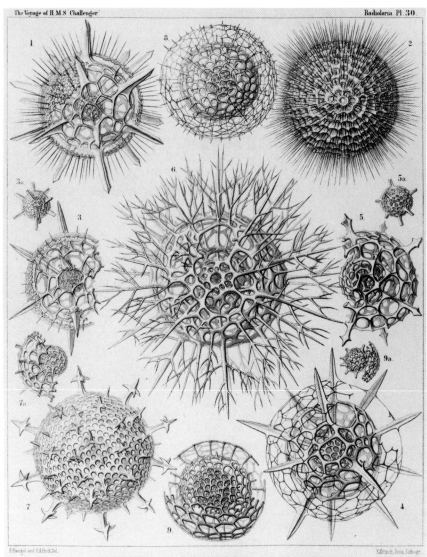

1—5 CROMYOMMA, 6.7 CROMYODRYMUS, 8.9 CROMYOSPHAERA.

Biomineralization

Cell Biology and Mineral Deposition

Kenneth Simkiss

Department of Pure and Applied Zoology
University of Reading
Reading, England

Karl M. Wilbur

Department of Zoology
Duke University
Durham, North Carolina

Academic Press, Inc.

Harcourt Brace Jovanovich, Publishers

San Diego   New York   Berkeley   Boston   London   Sydney   Tokyo   Toronto

The cover design brings together mineralized structures of three groups of organisms discussed in this volume: the algae, the sponges, and the molluscs. We are grateful to M. A. Borowitzka for the photograph showing coccoliths of the alga *Cyclococcolithus* (*lower left. See also* p. 67); to W. D. Hartman for the photograph of a sterraster of the sponge *Geodia* (*top. See also* p. 138); and to J. G. Carter for the photograph of calcitic prisms of the Cretaceous mollusc, *Inoceramus* (*lower right. See also* p. 240).

Academic Press, Inc.
San Diego, California 92101

United Kingdom Edition published by
Academic Press Limited
24–28 Oval Road, London NW1 7DX

Library of Congress Cataloging-in-Publication Data

Simkiss, Kenneth
  Biomineralization.

  Includes index.
  1. Biomineralization.   I. Wilbur, Karl M.
II. Title.
QH512.S58   1989   574.12'29   88-8002
ISBN   0-12-643830-7   (alk. paper)

Printed in the United States of America
89 90 91 92   9 8 7 6 5 4 3 2 1

To our wives,
Nancy and Jill

Contents

Part 3
Global Aspects

Preface

The past decade has seen a wealth of symposia and reviews dedicated to the understanding of the processes of biomineralization. This is because the phenomena cover interests as diverse as molecular biology and palaeontology. Most of these reviews are concerned with individual aspects of the subject. There is clearly a need, therefore, for a simple overview that indicates both the variety of organisms that are involved in these processes and the factual basis for the experiments that underlie our theoretical concepts. We hope that this book may contribute to this.

Our aim has been to bring into focus a diverse literature drawn from a wide range of taxa while at the same time posing a number of questions to the reader. To what extent do mechanisms of mineral deposition cross taxonomic lines? Do plants and microorganisms employ mineralizing processes that are distinct from those of animals? Can representatives of a single phylum deposit minerals by more than one process? These, of course, are the questions that comparative physiologists ask, and it could be argued that the answers will be provided by cell biologists, biochemists, and molecular biologists. At the moment, however, experimentalists in all three disciplines are finding it almost as difficult to explain the phenomena of biomineralization as did the microscopists of a century ago who first witnessed the elaborate skeletons of the radiolaria.

This book will not resolve these problems. We hope, however, that it will reveal a number of similarities in mechanisms while showing the variety of minerals that form in a vast array of organisms. Undoubtedly there are also differences, particularly in the ways that cells allocate their resources to this activity, transport ions, induce minerals, and manipulate their properties. Thus, differences as well as similarities have been looked for as opportunities for providing insights into mechanisms and opportunities for experimentation.

This survey of biomineralization is brief. Certain groups of organisms have been omitted simply because of the paucity of available information. Within the groups that are included, there has been no attempt to provide a thorough coverage of the literature. Rather, there has been a selection of papers and an attempt to pick out themes, and it is this, rather than a phylogenetic sequence, that has decided the arrangement of chapters.

We hope that the occasional reader will find an interest in some phase of biomineralization and that this, in turn, will lead to a further involvement. Within the discussion, we have therefore called attention to unsolved problems and aspects that need to be examined more closely. In discussing mechanisms we have also, on occasion, proposed explanations that still need testing. In examining these explanations, fundamental contributions to phenomena of biomineralization can still be made by the simplest observations as well as the most sophisticated apparatus.

The final form of this volume has come from the efforts of many friends and co-workers, to whom we are deeply grateful. They include Gill Bogue and Betty M. Bynum for typing the manuscript; Asenath M. Rasmussen for photography, preparing references, and permissions to authors and publishers; Mary McDonough-Gillan for assisting with typing changes in the manuscript, permissions, and references; Bertha Livingstone and Asenath M. Rasmussen for library searches; Drs. M. Okazaki, C.-M. Pan, and N. Watabe for unpublished data; Suzanne Ward for assistance with illustrations, references, index, and details in bringing the manuscript to its final form; Dr. S. C. Kunigelis for assistance with the chapter on molluscs; Dr. N. Watabe for many stimulating conversations and for reviewing and improving several chapters; and all those who provided original photographs of illustrations, including Drs. M. Borowitzka, J. Carter, and W. Hartman whose photographs were used in the cover design. We are grateful as well to the Department of Pure and Applied Zoology of the University of Reading and the Department of Zoology of Duke University for the generous contributions of their services throughout the preparation of this book and to the Duke University Research Council for financial assistance.

Kenneth Simkiss
Karl M. Wilbur

PART
1

Mechanisms

1

Biomineralization: The Discipline

Biomineralization, simply stated, is the process by which organisms convert ions in solution into solid minerals. To this incomplete definition one should add that the conversion is the result of cellular activities that make possible the necessary physicochemical changes for mineral formation and crystal growth. The end result is clearly apparent, for all manner of microorganisms, animals, and plants form a great variety of minerals. Some are delicate and beautifully sculptured while others impress by their massiveness, and most are closely related to very specific functions.

The subject of biomineralization is as varied as the range of organisms that form mineralized structures, and it spans a variety of disciplines. Biomineralization clearly holds an important place in the biological and medical studies of organisms that have mineralized skeletons, and it is obviously involved with the extracellular and intracellular events that occur in both normal and diseased states. Paleontology, so directly based on the resistance of mineralized portions of organisms to decomposition within geological strata, has a close tie with both the mineralizing process and the interpretation of its deposits. Most of the major kinds of skeletal materials appeared abruptly in 40 to 50 million years of the Cambrian. Since then only the corals, some algae, and the vertebrates have developed new skeletons in the marine habitat. The question may therefore be asked as to why these major developments in biomineralization are concentrated in about 1% of the Earth's history. In addition to these activities, biomineralization is also related to oceano-graphic and limnological research, for, as organisms convert ions from the external medium to mineral sediments, the chemistry of oceans and freshwaters will be changed. As sediments, reefs, and islands are formed from the mineral skeletons of marine organisms, the oceanic currents themselves are influenced. Thus, biomineralization has global implications, many of which are reflected in the concepts of biogeo-chemical cycling.

The study of biomineralization burgeoned in the 1960s and 1970s. One major reason for this expansion was the increased availability and use of transmission and scanning electron microscopes, which have continued to provide an abundance of information on the ultrastructure of the mineralized materials of animals, plants, and microorganisms (Lowen-stam, 1981). A second reason for the surge of interest has come from biochemical studies on membrane transport and the analysis of the or-ganic material associated with skeletal material. Both approaches fit di-rectly into the growing ideas of molecular biology and have received added impetus from the use of radioisotopes and the understanding of cell biology. Overviews of several areas of biomineralization research will be found in recent symposia (Watabe and Wilbur, 1976; Omori and Watabe, 1980; Simpson and Volcani, 1981; Nancollas, 1982; Westbroek and de Jong, 1983; and Leadbeater and Riding, 1986).

We shall start our survey by identifying the main aspects and phe-nomena associated with biomineralization.

I. Kinds of Biominerals and the Organisms That Form Them

The number of types of minerals of biological origin currently exceeds 60 (Table 1.1, 1.2) (Lowenstam and Weiner, 1983; see also Lowenstam, 1981; Wilson and Jones, 1984; Jones and Wilson, 1986), an increase of severalfold over the number reported in 1963 (Lowenstam, 1963). Of the biominerals reported in 1983, the majority have Ca as the major cation with Fe as the next most common metal. The most numerous biominerals grouped according to anions are phosphates, then oxides and carbonates. Roughly 25% of the biominerals are hydrated and 25% are amorphous, a fact that may be important when we come to consider the mechanisms of mineral formation. It will come as no surprise if the number of reported biominerals continues to increase.

Organisms with a capacity to form minerals are present among all the major groups from Monera to Chordata. There is a marked disparity, however, among taxa in the formation of individual minerals, as Table 1.1 shows. Minerals of Ca and Si are by far the most common in distribution among organisms. On the other hand, halides appear in very few groups, and sulfides and certain oxides have thus far been found only in the Monera. Oxalates and phosphates have a wider distribution; and the polymorphs of $CaCO_3$, aragonite, and calcite have the widest distribution of all. The tabulation of the types of minerals formed by the major groups in Table 1.1 is Monera, 9; other microorganisms, 10; fungi, 4; plants, 7; and invertebrates, 23.

II. Biological Functions of Minerals

Mineral deposition plays a major role in skeletal formation, and this function is evident in single-cell organisms as well as in many invertebrate phyla. The principal skeletal minerals are calcium carbonates, calcium phosphate, and silica (Table 1.2). In addition to the normal skeletal role, a considerable number of tissues appear to have their physical properties modified by the inclusion of metal deposits. The strongest and lightest carbonate and phosphate skeletons are well-ordered composites of organic polymers interspersed with a mineral phase. In others the mineral component is less distinct. Thus, a number of polychaete worms have jaws that are strengthened by copper and zinc deposits of uncertain composition (Bryan and Gibbs, 1980). A second and significant function of biomineral deposits is that they act as a storage system from which ions may be withdrawn during periods of special physiological demand. This occurs during the buffering of body fluids under both aerobic and anaerobic conditions, at the time of skeletal repair, when mineralization of an exoskeleton occurs following molting, and at times of reproductive stress and eggshell formation. Clearly, mineral deposits are not only available to the organism but they can be mobilized under a variety of stimuli.

Within the multicellular organisms, the capacity for mineralization depends primarily on specific tissues and not directly on the organism per se. This specificity is obvious, but it is worth emphasizing since it calls attention to cellular differences which dictate whether mineralization will or will not take place. Cellular specificity is nicely illustrated by those organisms in which one mineral is deposited by cells in one region

Table 1.1 Diversity and Phylum Distribution of Biomineralization Products in Extant Organisms[a]

Phylum	Monera	Protoctista															Fungi		Animalia											Plantae	
	Bacteria	Dinoflagellata	Haptophyta	Bacillariophyta	Phaeophyta	Rhodophyta	Chlorophyta	Zygnematophyta	Rhizopodea	Siphonophyta	Charophyta	Heliozoata	Radiolariata	Foraminifera	Mixomycota	Ciliophora	Basidiomycota	Deuteromycota	Porifera	Coelenterata	Platyhelminthes	Ectoprocta	Brachiopoda	Annelida	Mollusca	Arthropoda	Sipuncula	Echinodermata	Chordata	Bryophyta	Trachaephyta
Carbonates																															
Calcite	+	+	+			+				+	+			+					+	+	+	+	+	+	+	+	+	+	+	+	+
Aragonite	+	?				+	+			+				+					+	+		+		+	+	+	+		+		+
Vaterite					+	+	+																		+	+			+		+
Monohydrocalcite	+																												+		
Amorphous hydrated carbonate															+						+				+	+			+		
Phosphates																															
Dahllite																+		+							+				+		?
Francolite	+																						+						+		
$Ca_3Mg_3(PO_4)_4$																								+							
Brushite																									+						
Amorphous dahllite precursor																												+			
Amorphous brushite precursor																					+				+						
Amorphous whitlockite precursor																					+			+	+						
Amorphous hydrated ferric phosphate																								+	+			+			

Halides										
Fluorite						+	+			
Amorphous fluorite precursor						+	+			
Oxalates										
Whewellite	?	+	+			+		+	+	
Weddelite	?	+	+			+	+	+	+	
Sulfates										
Gypsum				+					?	+
Celestite		+							+	
Barite		+								
Silica										
Opal	?	+	+	+	+	+	+	+	+	
Fe-Oxides										
Magnetite	+	+	+			+	+	+	+	
Maghemite	?					+				
Goethite					+	+				
Lepidocrocite						+				
Ferrihydrite	+			+	+	+	+	+	+	+
Amorphous ferrihydrates					+	+	+		+	
Mn-Oxides										
"Todorokite"	+									
Fe-Sulfides										
Pyrite	+									
Hydrotroilite	+									

[a] From Mann (1983).

Table 1.2 The Types and Functions of the Main Inorganic Solids Found in
Biological Systems[a]

Mineral	Formula	Organism/function
Calcium carbonate		
Calcite	$CaCO_3$[b]	Algae/exoskeletons
		Trilobites/eye lens
Aragonite	$CaCO_3$	Fish/gravity device
		Molluscs/exoskeleton
Vaterite	$CaCO_3$	Ascidians/spicules
Amorphous	$CaCO_3 \cdot nH_2O$	Plants/Ca store
Calcium phosphate		
Hydroxyapatite	$Ca_{10}(PO_4)_6(OH)_2$	Vertebrates/endoskeletons, teeth, Ca store
Octacalcium phosphate	$Ca_8H_2(PO_4)_6$	Vertebrates/precursor phases in bone?
Amorphous	?	Mussels/Ca store
		Vertebrates/precursor phases in bone?
Calcium oxalate		
Whewellite	$CaC_2O_4 \cdot H_2O$	Plants/Ca store
Weddellite	$CaC_2O_4 \cdot 2H_2O$	Plants/Ca store
Group IIA metal sulfates		
Gypsum	$CaSO_4$	Jellyfish larvae/gravity device
Barite	$BASO_4$	Algae/gravity device
Celestite	$SrSO_4$	Acantharia/cellular support
Silicon dioxide		
Silica	$SiO_2 \cdot nH_2O$	Algae/exoskeletons
Iron oxides		
Magnetite	Fe_3O_4	Bacteria/magnetotaxis
		Chitons/teeth
Goethite	α-FeOOH	Limpets/teeth
Lepidocrocite	γ-FeOOH	Chitons (Mollusca)/teeth
Ferrihydrite	$5Fe_2O_3 \cdot 9H_2O$	Animals and plants/Fe storage proteins

[a] From Mann (1988).
[b] A range of magnesium-substituted calcites are also formed.

while a quite different mineral type is formed simultaneously by cells in
another part of the same organism. A familiar example is the deposition
of calcium carbonate in the shell of a mollusc by the mantle epithelium
while magnetite and SiO_2 are deposited by the epithelium which forms
the radular teeth (Lowenstam, 1962; Runham *et al.*, 1969). To add to the
subtlety of this system, one might add that, on occasion, even the same
cells may alternate at regular intervals to form different crystal orientations, as in the secretion of the crossed lamellar shells of some molluscs.
Clearly, the phenomena involved in mineral deposition vary as regards
both the cells that are involved and the products that they produce.

III. General Principles of Biomineralization

All cellular systems have three common properties. First, they consist largely of proteins dissolved in a salt solution and enclosed within lipid membranes. Second, for osmotic and other more subtle reasons the composition of the salts differs between the inside and outside of the lipid membranes. Third, in order to maintain osmotic stability they actively transport ions across these membranes. Among the divalent cations, magnesium is accumulated intracellularly while calcium is expelled across the cell membrane. Bicarbonate and phosphate ions are also often moved across these cell membranes and act as the main inorganic buffers of cells. In many organisms the soluble form of silica, i.e., silicic acid, $Si(OH)_4$, permeates the cell and is transported outward by metabolic processes associated with the cell membranes. Thus, the major components of the minerals shown in Table 1.1 are all involved in normal cell biology and share the properties that they are available, sparingly soluble and nontoxic in the solid state. It takes little imagination to realize that, given these properties, natural selection would soon favor those combinations which produce a variety of mineralized products, if these confer some advantage in survival. The membrane transport of ions and the formation of supersaturated solutions, the effects of proteins and polysaccharides on the formation and growth of crystals, and the enclosing of a space by membrane surfaces are the fundamental processes associated with most forms of biomineralization, and we shall return to each of these. As we study these processes in detail, we shall see that the ideas need to be extended and refined. The basic processes necessary for biomineralization reside, however, within most cells and it is the variety of their expression that provides the theme for this book on the study of biomineralization.

In summary, therefore, it appears that certain organisms of a given taxonomic group may deposit certain biominerals but not others. The deposits may be regarded as little more than stiffened cell walls or they may take the apparently brittle minerals and convert them into tough skeletons. Within the biomineralizing organism, the process of mineralization is restricted to specific tissues or organs. Here, then, are two kinds of specificity, one relating to control over the type of mineral deposited and the other to the metabolism of the tissue or organ that brings about mineral deposition and exploits its properties. The systems that determine these two kinds of specificity are unknown but, from studies of the mechanisms employed by various taxa, two aspects of biomineralization may be considered. These are, first, whether or not the mechanisms necessary to form a single mineral type are similar in different taxa and, second, whether or not different mineral types are deposited and used by systems with common properties. The diversity of the information that is available to answer these two questions will occupy much of this book.

References

Bryan, G. W., and Gibbs, P. E. (1980). Metals in neired polychaetes: The contribution of metals in the jaws to the total body burden. *J. Mar. Biol. Assoc. U.K.* **60,** 641–654.

Jones, D., and Wilson, M. J. (1986). Biomineralization in crustose lichens. *Syst. Assoc. Spec. Vol.* **30,** 91–105.

Leadbeater, B. S. C., and Riding, R. (eds.) (1986). *Syst. Assoc. Spec. Vol.* **30,** 1–401.

Lowenstam, H. A. (1962). Goethite in radular teeth of recent marine gastropods. *Science* **137,** 279–280.

Lowenstam, H. A. (1963). Biological problems relating to the composition and diagenesis of sediments. *In* "The Earth Sciences. Problems and Progress in Current Research" (T. W. Donnelly, ed.), pp. 137–195. Univ. of Chicago Press, Chicago.

Lowenstam, H. A. (1981). Minerals formed by organisms. *Science* **211,** 1126–1131.

Lowenstam, H. A., and Weiner, S. (1983). Mineralization by organisms and the evolution of biomineralization. *In* "Biomineralization and Biological Metal Accumulation" (P. Westbroek and E. W. de Jong, eds.), pp. 191–203. Riedel, Dordrecht, The Netherlands.

Mann, S. (1983) Mineralization in biological systems. *Struct. Bonding, Berlin* **54,** 125–174.

Mann, S. (1988). Molecular recognition in biomineralization. *Nature (London)* **332,** 119–124.

Nancollas, G. H. (ed.) (1982). "Biological Mineralization and Demineralization," pp. 1–415. Springer-Verlag, Berlin.

Omori, M., and Watabe, N. (eds.) (1980). "The Mechanisms of Biomineralization in Animals and Plants," pp. 1–310. Tokai Univ. Press, Tokyo, Japan.

Runham, N. W., Thornton, P. R., Shaw, D. A., and Wayte, R. C. (1969). The mineralization and hardness of the radular teeth of the limpet *Patella vulgata* L. *Z. Zellforsch. Mikrosk. Anat.* **99,** 608–626.

Simpson, T. L. and Volcani, B. E. (eds.) (1981). "Silicon and Siliceous Structures in Biological Systems." Springer-Verlag, Berlin.

Watabe, N., and Wilbur, K. M. (eds.) (1976). "The Mechanisms of Mineralization in the Invertebrates and Plants," pp. 1–461. Univ. of South Carolina Press, Columbia, South Carolina.

Westbroek, P., and de Jong, E. W. (eds.) (1983). "Biomineralization and Biological Metal Accumulation," pp. 1–533. Riedel, Dordrecht, The Netherlands.

Wilson, M. J., and Jones, D. (1984). The occurrence & significance of manganese oxalate in *Pertusaria corallina* (Lichens). *Pedobiologia* **26,** 373–379.

2

The Deposition of Minerals

Organisms, in forming minerals, create microenvironments in which solids of a specific kind often grow to a particular size. In many invertebrates, these solids become units of the skeleton and show a species-specific morphology. As we have already mentioned, some of these minerals, such as silica, are amorphous, by which we mean that they lack the long-range order of regular repeating units which characterize a crystal. The formation of such materials is, however, somewhat more difficult to understand, so we will initially concern ourselves with crystals whose properties are more easily quantified.

The starting point for crystal deposition is the supersaturation of a fluid with the subsequent nucleation and growth of a specific mineral within it. The detailed ionic composition and the inorganic and organic content of the microenvironments in which crystals form is quite complex and to a considerable degree unknown except in certain plant vacuoles. Despite this, the basic mechanisms of crystal formation in simple systems are so well characterized that we can apply them to organisms with considerable confidence.

I. Fluid Supersaturation

Crystals will only form from solutions which contain the relevant ions if their concentrations exceed the solubility product constant. Usually, crystal formation only proceeds in a predictable way if there is also a crystallization center in the solution. Thus, technically, the solubility product is defined as the thermodynamic product of the activities of the ions in a solution that is in equilibrium with a pure solid. The solubility product constant (K_{sp}) can be expressed

$$K_{sp} = [C^+][A^-](f_{C^+} f_{A^-})$$

in which $[C^+]$ and $[A^-]$ are ion concentrations and $f_{C^+} f_{A^-}$ is the square of the mean activity coefficient of the ions (Clark, 1948). This formulation assumes a pure inorganic solution containing only those ions which will form the mineral. Put more simply, this means that in order to form a crystal of, say, calcite ($CaCO_3$), then the product of the concentrations of calcium and carbonate ions must exceed a certain value. Note, however, two aspects of this law. First, it is the *product* of the ions that is important. If the activity of only one ion is increased it will cause a crystal to form as soon as the product of the two ions is exceeded. In many biological systems it is therefore argued that if the calcium level is increased by the transport of that ion, then that may be all that is necessary to start crystallization. Alternatively, if the carbonate ion activity is increased by fixing carbon dioxide during photosynthesis in a bicarbonate-containing solution ($2HCO_3^- \rightarrow CO_2 + H_2O + CO_3^{2-}$), that also could be the driving force to initiate calcification. It is not necessary to increase the concentrations of both ions simultaneously to initiate biomineralization.

The second important aspect of the concept of solubility product constants is that they are specific for a particular arrangement of ions in a crystal. Calcite and aragonite are both pure forms of calcium carbonate but each has its own solubility product constant. For calcite the value is approximately 4.7×10^{-9}, while for aragonite it is 6.9×10^{-9} $kmol^2$ m^{-3}. Aragonite is, therefore, more soluble than calcite. As such, one might expect it to be less common. If, however, you slowly add calcium to seawater it is aragonite, not calcite, that crystallizes out of solution. The explanation of this apparent contradiction is that the magnesium ion in the seawater interferes with the crystallization process and favors the more soluble polymorph. In living systems, various ions and organic molecules will always be present and some of these will probably be in contact with crystal surfaces and may be inhibitory. Because of the effects of such extraneous ions and molecules and their changes in concentrations, calculations of the solubility product constant are difficult to perform and not always meaningful in many biological systems. So, we shall simply accept the physicochemical fact that, in mineralizing organisms, ions must be present at concentrations which will exceed the solubility product constant and that when this occurs it will result in the deposition of a solid mineral phase.

II. Crystal Nucleation

In understanding the sequence of changes in the formation of crystals, it is useful to return to the simple condition of a pure supersaturated solution. As the concentrations of the ions in solution increase, the ions begin to associate into small clusters but these are unstable and dissociate again because of their relatively large surface area in relation to their size. If the clusters are to grow and become more stable nuclei, there must be an energy input to create a new solid interface with the fluid phase. This energy is derived as bonds are formed in the solid phase

Figure 2.1 Free energy of an atomic cluster in relation to its size (r). G_N^* refers to a given value of the Gibbs free energy of the cluster ΔG_N^*, r is the radius of spherical clusters, r^* refers to a given supersaturation ratio, σ is the surface free energy per unit surface area, ΔG_v is the free energy change per mole associated with liquid–solid phase change, and V_m is the molar volume. (From Garside, 1982.)

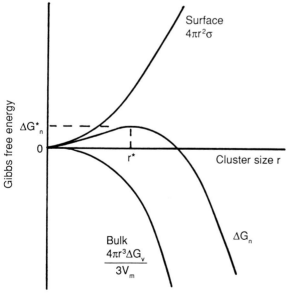

(Mann, 1983). The required free energy for the formation of a nucleus, G_N, is expressed by

$$G_N = G_{surface} + G_{solid}$$

in which $G_{surface}$ is the free energy change at the solid–liquid interface and G_{solid} is the negative free energy change in the energy released by bond formation in the solid phase. As the ion clusters increase in size, G_N increases to a maximum (Fig. 2.1) at which the nuclear size is said to be critical (Garside, 1982; Mann, 1983). The number of ions in a nucleus of critical size is of the order of 10–1000. Clusters smaller than the critical size tend to dissolve, so decreasing their free energy. Above the critical size, growth of nuclei is favored in that this also results in a free energy decrease (Fig. 2.1).

For a given period of time, the number of nuclei which form in a unit volume of solution will be a function of the level of supersaturation (Garside, 1982). At levels below a given value (S^*), the nucleation rate is so low that crystals do not occur (Fig. 2.2). Within the range where nucleation is possible, the solution is said to be *metastable.* As supersatu-

Figure 2.2 Form of the primary nucleation rate equation for homogeneous and heterogeneous nucleation. Note the relationship between the supersaturation of the solution and the rate of crystal nucleation. (From Garside, 1982.)

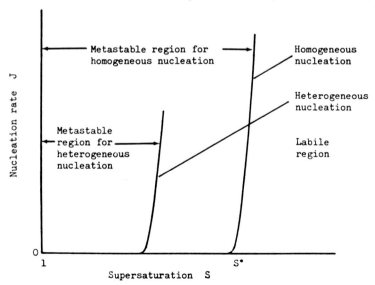

ration is increased above S^* or, alternatively, if crystals are added to the solution (seeding), the rate of formation of nuclei is markedly increased.

Homogeneous or spontaneous nucleation of the type we have been describing probably never occurs in biological mineralizing systems in that other ions and molecules and various bodies and surfaces are always present and probably influence the process (see Watabe, 1981). Under these conditions, *heterogeneous nucleation* occurs and mineral is deposited on existing surfaces by a process that is in fact capable of being maintained at lower supersaturations than homogeneous nucleation (Fig. 2.2). The lower level is due to the fact that there is a reduction in the free energy change (G_N) when clusters form in contact with preexisting surfaces. Heterogeneous nucleation is clearly an advantageous mechanism for organisms in that it simultaneously requires less metabolic energy for the active transport of ions while providing a mechanism for controlling the sites of mineralization. Such a system of heterogeneous nucleation is normally achieved in biological systems by using the organic materials that are associated with most sites of mineral deposition.

The time required for nuclei to grow into crystals in a supersaturated solution can be measured either because it causes a change in the optical properties of the solution or because it can be detected by sampling the solution at intervals and collecting the crystals. The time of first appearance of crystals in a solution is designated the *induction time*. The phenomenon can also be measured as the time required for the first noticeable change in the composition of the solution, although this can usually only be detected after the optical change has occurred (Garside, 1982). These methods have proved useful in quantifying which substances inhibit nucleation and crystal growth of calcium phosphate (Termine and Conn, 1978) and calcium carbonate (Borman *et al.*, 1982; Sikes and Wheeler, 1983; Wilbur and Bernhardt, 1984).

III. Crystals and Their Growth

Later in our discussion of mineral deposition in organisms, we shall find it convenient to recall a few basic concepts of crystals and their growth. We introduce them here.

Nuclei, whose formation we have just considered, continue to add ions onto their surfaces as they grow to form crystals. We say that they have become crystals when they have a structure of identical units in a repeated pattern in three dimensions. Figure 2.3 shows such a three-dimensional array of a very small portion of $CaCO_3$ atoms in their spatial

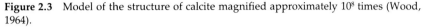

Figure 2.3 Model of the structure of calcite magnified approximately 10^8 times (Wood, 1964).

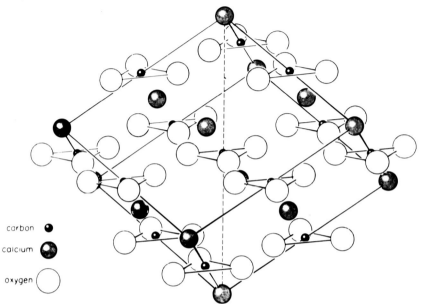

carbon

calcium

oxygen

arrangement in calcite (Wood, 1964). The distances between the atoms are magnified approximately 10^8 times. Each carbon atom is surrounded by three oxygen atoms. This pattern together with the calcium atoms repeats and repeats throughout the crystal. The arrangements of the same atoms in vaterite and aragonite, two other types of $CaCO_3$, are somewhat different. We therefore speak of calcium carbonate as being polymorphic since the same atoms can exist in several crystalline forms with the different stabilities we have already mentioned.

Because of their orderly three-dimensional atomic arrangements, the mineral crystals deposited by organisms can be identified by *X-ray diffraction*. For this determination, one or more crystals are exposed to an X-ray beam. The X-rays are displaced by the atoms in the crystal so that a particular diffraction pattern is formed and this can subsequently be analyzed to identify the crystal type. Not infrequently, it is necessary to identify crystals within cells that are being studied at the ultrastructural level. In this case, crystallographic analysis can often be carried out in the electron microscope by *electron diffraction*. The pattern resulting from the diffraction of the electron beam by the crystal is recorded on film and subsequently analyzed in a similar way to X-ray patterns. An example of an electron diffraction pattern is shown in Fig. 2.4. The structure shown is a very small fragment of a piece of $CaCO_3$ called a coccolith, formed in a Golgi vesicle of the alga *Emiliania huxleyi*. With high resolution electron microscopes, it is also now possible to obtain visual imaging of the crystal lattice. This provides the possibility for direct studies of the nucleation of biominerals and is a most exciting development (Mann, 1986).

The growth of crystals occurs by the addition of ions to the crystal faces in a process that is not necessarily uniform. As a result, a crystal face develops dislocations and steps which are important for subsequent growth (Nielsen and Christoffersen, 1982; Mann, 1983). Thus, as crystal growth continues, minute crystals may form on the smooth face, at the edge of a step, or at a so-called kink in a step (Fig. 2.5). In fact the growth of a crystal is more likely at a kink or edge than anywhere else since these sites are particularly favorable energetically (Figs. 2.6 and 2.7). For this reason, the kinks tend to fill in and the steps grow towards a uniformly shaped crystal (Fig. 2.5).

Figure 2.4   (A) Fragment of a coccolith of *Emiliania huxleyi*. x63,200. (B) Electron diffraction pattern of the fragment. Axes of crystals are indicated by arrows. (Watabe, 1967.)

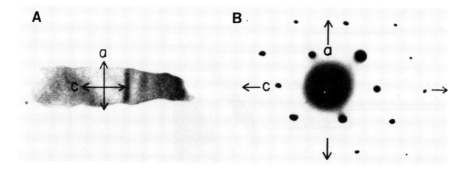

tion, the phenomenon becomes increasingly under the control of the organism. The nature of these controls remains as one of the central problems of biomineralization and we will therefore initially consider those organisms where it is often least well developed.

References

Borman, A. H., de Jong, E. W., Huizinga, M., Kok, D. J., Westbroek, P., and Bosch, L. (1982). The role of calcium carbonate crystallization of an acid Ca^{2+}-binding polysaccharide associated with coccoliths of *Emiliania huxleyi*. *Eur. J. Biochem.* **129**, 179–183.

Burton, W. K., Cabrera, N., and Frank, F. C. (1951). The growth of crystals and the equilibrium structure of their surfaces. *Philos. Trans. R. Soc. London, Ser. A* **243**, 299–358.

Clark, W. M. (1948). "Topics in Physical Chemistry." Williams & Wilkins, Baltimore, Maryland.

Garside, J. (1982). Nucleation. *In* "Biological Minerlization and Demineralization" (G. H. Nancollas, ed.), pp. 23–25. Springer-Verlag, Berlin.

Grigor'ev, D. P. (1965). "Ontogeny of Minerals." Israel Program for Scientific Translations, Jerusalem.

Mann, S. (1983). Mineralization in biological systems. *Struct. Bonding (Berlin)* **54**, 126–174.

Mann, S. (1986). The study of biominerals by high resolution transmission electron microscopy. *Scanning Electron Microscopy* **11**, 393–413

Nielsen, A. E., and Christoffersen, J. (1982). The mechanisms of crystal growth and dissolution. *In* "Biological Mineralization and Demineralization" (G. H. Nacollas, ed.), pp. 37–77. Springer-Verlag, Berlin.

Sikes, C. S., and Wheeler, A. P. (1983). A systematic approach to some fundamental questions of carbonate calcification. *In* "Biomineralization and Biological Metal Accumulation" pp. 285–289. Riedel, Dordrecht, The Netherlands.

Termine, J. D., and Conn, K. M. (1976). Inhibition of apatite formation by phosphorylated metabolites and macromolecules. *Calcif. Tissue Res.* **22**, 149–157.

Watabe, N. (1967). Crystallographic analysis of the coccolith of *Coccolithus huxleyi*. *Calcif. Tissue Res.* **1**, 114–121.

Watabe, N. (1981). Crystal growth of calcium carbonate in the invertebrates. *Prog. Cryst. Growth Charact.* **4**, 99–147

Wilbur, K. M. and Bernhardt, A. M. (1984). Effects of amino acids, magnesium, and molluscan extrapallial fluid on crystallization of calcium carbonate: *In vitro* experiments. *Biol. Bull. (Woods Hole, Mass.)* **166**, 251–259.

Wood, E. A. (1964). "Crystals and Light." Van Nostrand, Princeton, New Jersey.

3

The Origins
of Biomineralization—
Microbial Systems

I. Biotic and Abiotic Interactions

The composition of the atmosphere that was emitted from the interior of the Earth at the time that it was formed is not known, but it was almost certainly reducing and contained little or no oxygen. It was within these conditions that the bacteria initially evolved and photosynthesis, or the ability to synthesize organic molecules from light and simple inorganic compounds, was one of the fundamental biochemical pathways. The primary reaction is

$$CO_2 + 2H_2A \xrightarrow{\text{light}} (CH_2O) + H_2O + 2A \qquad (3.1)$$

where H_2A is an electron donor that becomes oxidized in the reaction. In the so-called photosystem I that survives in the green and purple bacteria, electrons are accepted from compounds such as H_2A, provided that they release the electrons at relatively high energy levels. Thus, A in Eq. (3.1) may be sulfur as H_2S which becomes oxidized to S during the photosynthetic process (Krumbein, 1979). When, however, the cyanobacteria or blue-green algae evolved, they incorporated additional reactions (photosystem II) into these biochemical pathways so that it was then possible to release electrons from the much lower energy levels of water. H_2A in Eq. (3.1) then became H_2O and oxygen was emitted during the photosynthetic formation of carbohydrate.

$$CO_2 + H_2O \xrightarrow{\text{light}} (CH_2O) + O_2 \tag{3.2}$$

With the advent of the resulting oxidizing atmosphere there was a major shift in the composition of the inorganic environment. Iron, for example, which had become a required element for all forms of life, suddenly passed from the relatively soluble form of Fe^{2+} [which has a solubility product constant for $Fe(OH)_2$ of 4.79×10^{-17}] to the highly insoluble form of Fe^{3+} [which has a solubility product for $Fe(OH)_3$ of 2.67×10^{-39}). Although iron accounts for roughly 5% of the mass of the Earth's crust, its insolubility in an oxidizing atmosphere rendered it biologically scarce. However, bacteria were able to obtain it by secreting chelating agents or siderophores.

II. Bacterial Mineralization in the Environment

Because bacteria manipulate such a wide variety of biochemical pathways and because they frequently release a variety of metabolic products to the environment, they are often the center of a wide range of mineralization processes. The photosynthetic removal of carbon dioxide from a solution such as freshwater or seawater that contains calcium and bicarbonate ions may lead to the precipitation of calcium carbonate by driving the following reaction:

$$Ca^{2+} + 2HCO_3^- \rightarrow CaCO_3 \text{ (prec)} + H_2O + CO_2 \text{ (fixed)} \tag{3.3}$$

Carbon dioxide fixation may also occur in the dark by a variety of chemilithotropic organisms such as nitrifying bacteria or iron-oxidizing bacteria. It is also carried out by the nonoxygenic photosynthesizing green and purple bacteria and by the oxygen-emitting cyanobacteria. In a suitable medium, therefore, a number of soil bacteria will form calcite or, in the presence of magnesium ions, aragonite (Greenfield, 1963). The fixation of carbon dioxide is, however, only one of the very diverse range of bacterial metabolic activities which results in biomineralization. Nitrate reduction by heterotrophs may also induce calcium carbonate deposition:

$$Ca(NO_3)_2 + 3H_2 + C \text{ (organic)} \rightarrow CaCO_3 + 3H_2O + N_2 \tag{3.4}$$

as may sulfate reduction, e.g., by *Desulfovibrio*:

$$CaSO_4 + 2(CH_2O) \text{ (organic)} \rightarrow CaCO_3 + H_2S + CO_2 + H_2O \tag{3.5}$$

In the presence of ferrous ions, this could result in the deposition of ferrous sulfide as well as calcium carbonate, i.e.,

$$CaSO_4 + 2(CH_2O) + Fe \rightarrow CaCO_3 + FeS + CO_2 + H_2O \qquad (3.6)$$

which emphasizes precipitation and the somewhat indiscriminate nature of these types of mineralization (Trudinger, 1979).

A somewhat similar effect may be produced by both aerobic and anaerobic bacteria which release ammonia following the oxidative deamination of amino acids, i.e.,

$$2NH_4OH + Ca(HCO_3)_2 \rightarrow CaCO_3 + (NH_4)_2CO_3 + 2H_2O \qquad (3.7)$$

or

$$(NH_4)_2CO_3 + CaSO_4 \rightarrow CaCO_3 + (NH_4)_2SO_4 \qquad (3.8)$$

Krumbein (1979) has discussed these mechanisms at length. According to Deelman (1975), it is, however, possible to identify two types of mechanism in bacterial carbonate deposition, namely, those that result from metabolic pathways of the type we have been discussing in relation to CO_2 fixation, and those which result from a change in the redox potential. In this latter case at a pH of about 7 and with a redox potential (E_o) of 250 to 300 mV, bicarbonate combines with H^+ to form methane and calcium carbonate, i.e.,

$$3Ca^{2+} + 4HCO_3 + 6H^+ + 8e \rightarrow CH_4 + 3H_2O + 3CaCO_3 \qquad (3.9)$$

Redox potentials also influence the metabolism of metals such as iron and manganese, which can occur as soluble reduced forms (Mn^{2+}, Fe^{2+}) or insoluble oxidized forms (Mn^{4+}, Fe^{3+}). Thus bacterial activities which affect pH or E_o may also influence the distribution of these metals indirectly via redox states or in a very direct way by providing inorganic binding sites.

From this discussion it will be apparent that a large number of bacterial activities are likely to induce mineral formation if they occur in suitable solutions. It must be emphasized, however, that most of these products occur extracellularly and there will only be a deposition of mineral if the solubility product is exceeded at that local site. Many of the most spectacular consequences of bacterial biomineralization are only apparent, therefore, in the rather complex ecological aggregates of organisms that permit these local environments. This is perhaps best seen in stromatolites, which are biosedimentary structures built by microbial trapping and binding and/or precipitating of sediments. Stromatolites are well known in the fossil record and still occur in a few limited sites such as Western Australia. In effect, the stromatolite is a microbial community, although it may be dominated by one species such as the cyanobacterium *Entophysalis major*. Mats of these organisms contain organic sheaths and may recycle inorganic elements within the community. The mineralization of the community starts during the summer, when, as a result of algal photosynthesis, the pH rises and the highest concentrations of $CaCO_3$ occur in the surrounding sea. Mineral deposition is initially in the polysaccharide coat around the bacteria, although it rapidly permeates this entire gel as an amorphous calcareous deposit. As the mat hardens it loses its translucent appearance, and becomes brown and opaque with a hard interior. Simultaneously there is a deterioration of the *Entophysalis* and only a thin film of living cells remains on the outside of the mineral deposit. This film mat is then populated by a variety of bacterial and algal forms, some of which are stromatolite-destroying as opposed to stromatolite-forming (Golubic, 1983), and the subsequent fate of the deposit depends on the interplay of these diverse forms.

Much more complex ecological situations have obviously occurred with bacteria interacting to induce large mineral deposits. One such community in Laguna Figueroa, Mexico, is described by Margulis and Stolz (1983). From the surface this evaporite flat looks desolate and uninhabited, but beneath the mineral crust is a diverse microbial community. The upper half centimeter contains a photosynthetic community which fixes carbon dioxide and enhances aragonite deposition. Beneath this are several heterotrophic species of bacteria which deposit manganese oxides while sulfate-reducing bacteria in the anaerobic muds produce hydrogen sulfide and precipitate a number of metal sulfides. Such complex ecological and mineralogical interactions have probably been of great importance in the evolution of the Earth's surface but it should be emphasized that they represent only the first phase of the biomineralization process, namely, the interaction of metabolic pathways with environmental materials. There is little evidence among the bacteria for the extracellular *organization* of the mineral deposits which is to become an important feature of biomineralization among the eukaryotes. Mineralization may, however, occur in association with the outer coverings of bacteria. The major components of these coats or envelopes are peptidoglycans, proteins, and polysaccharides. The acidic groups of the peptidoglycans and acidic polysaccharides can bind metal ions and, in this manner, mineral formation may be initiated in certain species. An example of extracellular mineralization of this type is shown in Fig. 3.1 in which ferric hydroxide is deposited in the envelope of *Leptothrix pseudovasculata* (Caldwell and Caldwell, 1980).

Manganese-oxidizing bacteria, known from most environments, form manganates (Nealson, 1983), but the mechanism of oxidation is unclear.

Figure 3.1 A transmission electron micrograph of a *Leptothrix* sp. collected from iron-containing pools in the Cave of the Mounds Caverns, which is located in central Wisconsin. The cell filaments are surrounded by a sheath encrusted with a fibrillar matrix of ferric hydroxide (a). This matrix of fibrils is also observed within the cells (b). Bar represents 1.04 μm (Caldwell and Caldwell, 1980).

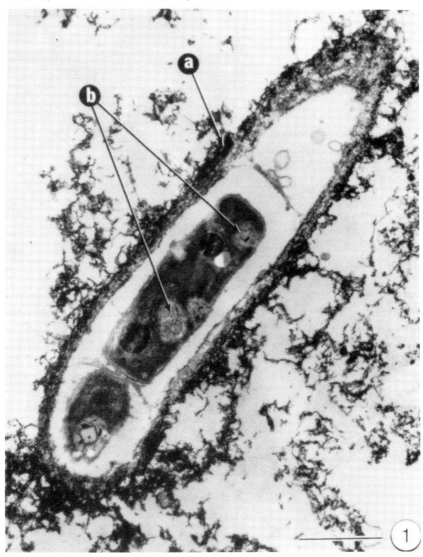

Extracellular acidic polysaccharides can be isolated from *Pedomicrolium* sp. and these bind Mn^{2+} and hasten its oxidation. Spore coat proteins from marine *Bacillus* sp. also oxidize Mn^{2+} with no energetic advantage of the organism. The possibility that manganese might serve as an energy source for bacteria has fascinated workers for many years, but to date there is no unequivocal evidence to support this. What is clear, however, is that, by drawing energy and metabolites from the environment, even simple forms of life like bacteria induce the formation of mineral deposits. This, then, is one way in which biomineralization started, but there are other ways with even greater effects.

III. Bioaccumulation and Intracellular Mineralization

The energy that life depends on is manipulated and released by oxidation–reduction reactions and is subsequently captured by phosphorylation of organic compounds. All forms of life depend on iron for some aspect of their oxidation–reduction chemistry and thus accumulate this potential source of mineralization. They also depend on stores of phosphate for use in their metabolism. It may be that this phosphate based metabolism necessitated the exclusion of calcium from cells since calcium phosphate is a very insoluble product. Whether this is the reason or not, it appears that all cells pump calcium out of their cytoplasm into the extracellular environment and replace it with magnesium. Frequently, one of the first signs of a pathological disturbance is the leaking of calcium into a cell and the precipitation of intracellular minerals. Some cells do, however, allow calcium to enter the cytoplasm and in-

Table 3.1 Types of Minerals Produced by Microorganisms

Minerals	Species	References
Extracellular iron sulfides and ferric hydroxides	*Leptothrix pseudovacuolata*	Caldwell and Caldwell (1980)
Intracellular magnetite particles	Several	Blakemore *et al.* (1980); Blakemore (1982)
Intracellular apatite	*Bacterionema matruchotii*	Ennever *et al.* (1971); Boyan *et al.* (1984)
Intracellular apatite	*Escherichia coli*	Ennever *et al.* (1974)
Intracellular apatite	*Streptococcus mutans*	Ennever *et al.* (1972)
Intracellular apatite	*Enterobacter cloacae*	Ennever *et al.* (1972)
Intracellular apatite	*Algaligenes marshalli*	Ennever *et al.* (1972)
Intracellular apatite	*Proteus mirabilis*	Keefe and Smith (1977)
Intracellular apatite	*Staphylococcus epidermidis*	Boyan *et al.* (1984)

stead of pumping it out against a large concentration gradient, they compartmentalize it and render it insoluble in some specialized area of the cytoplasm. Several species of bacteria of dental and medical interest deposit hydroxyapatite intracellularly when cultured in media containing calcium and phosphate (Table 3.1) (Streckfuss *et al.*, 1974; Ennever *et al.*, 1981). *Bacterionema matruchotii*, an oral microorganism, has been used extensively under well-defined culture conditions in an examination of the mechanism of mineral deposition (Boyan *et al.*, 1984). The mineral is formed in this species only after several days of culture (Ennever and Takazoe, 1973; Boyan *et al.*, 1984). The crystals of hydroxyapatite may be scattered in the cytoplasm in light mineralization, or in heavy deposition they may occupy most of the cell interior (Fig. 3.2) (Takazoe *et al.*, 1963).

Figure 3.2   Apatite crystals in *Bacterionema matruchotii* ATCC 14266. (Courtesy of J. Ennever.)

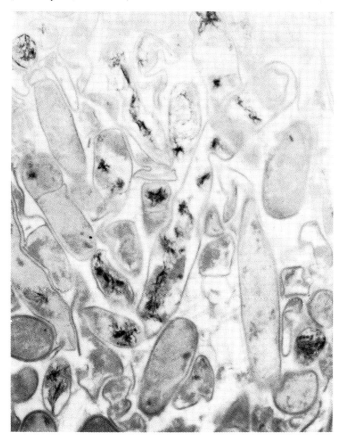

During the period before mineralization occurs, the proteolipid concentration increases, its inorganic phosphate (P_i) and ionic composition changes, and the Ca:Mg:P_i ratio is shifted. At the time of mineralization, it is not at all certain that the mineralized cells are alive. In fact, it is known that dead cells of this species will form hydroxyapatite intracellularly (Ennever and Takazoe, 1973). In dead cells, the cell membrane would permit the free entrance of mineral ions and the establishment of a new equilibrium between the medium and the cytosol. The result would be an increase in the internal concentration of calcium and phosphate (Boyan et al., 1984). It appears that the deposition of hydroxyapatite may be initiated by a lipoprotein which can be extracted from B. matruchotii and the same material will bring about hydroxyapatite crystallization in a metastable solution (Ennever et al., 1978). In the process of apatite formation within the cells, protein–lipid complexes are presumed to induce the formation of calcium–phospholipid–phosphate complexes (Boyan et al., 1984). We see, therefore, in the bacteria the start of intracellular mineralization and the role of organic molecules in this process.

More recently, however, an exciting form of iron mineralization has also been discovered in bacteria. Various species of bacteria found in natural waters exhibit magnetotactic behavior, that is, their direction of movement is determined by an orientation corresponding to the lines of a magnetic or geomagnetic field. This intriguing behavior is the result of particles of magnetite (Fe_3O_4) within the cell (Blakemore, 1975, 1982; Frankel et al., 1978; Balkwill et al., 1980; Bazylinski et al., 1988). The cells move toward the north if the north-seeking pole of the magnetic particles is oriented forward and toward the south with the reverse orientation of the particles. The particles may be in the form of cubes, rectangular parallelepipeds, or arrowheads (Frankel and Blakemore, 1984) and are arranged in one or two chains (Fig. 3.3) (Towe and Moench, 1981).

In order to determine the state of the iron in the particles, the bacterium Aquaspirillum magnetotacticum was labeled with ^{57}Fe and studied using Mossbauer spectroscopy with the following results (Frankel and Blakemore, 1984). Iron chelated to quinic acid enters the cell as Fe^{3+}. A sequence of changes in the state of the iron can then be followed and was found to be Fe^{3+} quinate $\rightarrow$ Fe^{2+} $\rightarrow$ low-density hydrous ferric oxide $\rightarrow$ ferric hydrite $\rightarrow$ Fe_3O_4. The unique step in this sequence appears to be the conversion of the final precursors into magnetite. The other stages in the reaction are found in many cells which use the process to fill the iron storage molecule, ferritin. This indicates how even minor alterations to normal metabolic pathways can be used by organ-

Figure 3.3 Stained ultrathin section of a magnetic coccus showing magnetite grains along the cell wall, each enclosed within an intracytoplasmic membrane. ×137,000. (Towe and Moench, 1981).

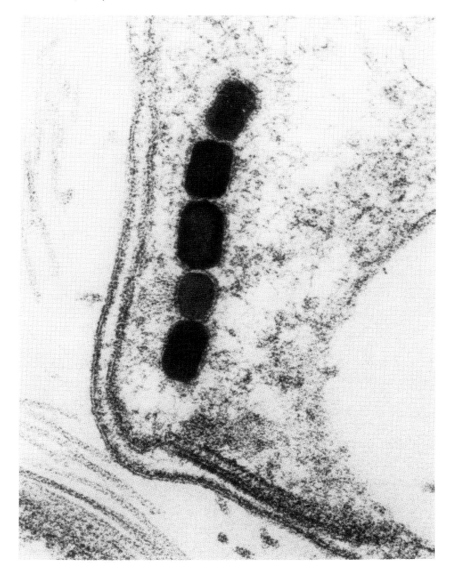

isms to induce mineralization. There is, however, an additional principle to be recognized in this discovery. The magnetite deposits that are formed are single magnetic domains. They may be formed within organic membranes which appear to stabilize the growth of particular crystal faces, although this has not yet been demonstrated with certainty. If this is confirmed, then it may be an example of epitaxy, for the magnetite crystals are highly ordered with a characteristic morphology.

Bacteria have had a fundamental influence on the evolution of other forms of life and on the composition of the Earth's crust. Their interaction with the environment has often led indirectly to the deposition of minerals but it is quite clear that they have also started to evolve the spatial, chemical, and structural processes that form the basis of all forms of biomineralization.

References

Balkwill, D. L., Maratea, D., and Blakemore, R. P. (1980). Ultrastructure of a magnetotactic spirillium. *J. Bacteriol.* **141**, 1399–1408.

Bazylinski, D. A., Frankel, R. B., and Jannasch, H. W. (1988). Anaerobic magnetite production by a marine, magnetotactic bacterium. *Nature (London)* **334**, 518–519.

Blakemore, R. P. (1975). Magnetotactic bacteria. *Science* **190**, 377–379.

Blakemore, R. P. (1982). Magnetotactic bacteria. *Annu. Rev. Microbiol.* **36**, 217–238.

Blackmore, R. P., Frankel, R. B., and Kalmijn, A. J. (1980). South-seeking magnetotactic bacteria in the Southern Hemisphere. *Nature (London)* **286**, 384–385.

Boyan, B. D., Landis, W. J., Knight, J., Dereszewski, G., and Zeagler, J. (1984). Microbial hydroxyapatite formation as a model of proteolipid-dependent membrane-mediated calcification. *Scanning Electron Microsc.* **4**, 1793–1800.

Caldwell, D. E., and Caldwell, S. J. (1980). Fine structure of *in situ* microbial iron deposits. *Geomicrobiol. J.* **2**, 39–53.

Deelman, J. C. (1975). Two mechanisms of microbial carbonate precipitation. *Naturwissenschaften* **62**, 484–485.

Ennever, J., and Takazoe, I. (1973). Bacterial calcification. In "Biological Mineralization" (I. Zipkin, ed.), pp. 629–648. Wiley (Interscience), New York.

Ennever, J., Vogel, J. J., and Streckfuss, J. L. (1971). Synthetic medium for calcification of *Bacterionema matruchotti*. *J. Dent. Res.* **49**, 1327–1330.

Ennever, J., Vogel, J. J., and Brown L. R. (1972). Survey of microorganisms for calcification in a synthetic medium. *J. Dent. Res.* **51**, 1483–1486.

Ennever, J., Vogel, J. J., and Streckfuss, J. L. (1974). Calcification by *Escherichia coli*. *J. Bacteriol.* **119**, 1061–1062.

Ennever, J., Riggan, L. J., Vogel, J. J., and Boyan-Salyers, B. (1978). Characterization of *Bacterionema matruchotii* calcification nucleator. *J. Dent. Res.* **57**, 637–642.

Ennever, J., Streckfuss, J. L., and Goldschmidt, M. L. (1981). Calcifiability comparison among selected microorganisms. *J. Dent. Res.* **60**, 1793–1796.

Frankel, R. B., and Blakemore, R. P. (1984). Precipitation of Fe₃O₄ in magnetotactic bacteria. *Philos. Trans. R. Soc. London, Ser. B* **304**, 567–574.

Frankel, R. B., Blakemore, R. P., and Wolfe, R. S. (1978). Magnetite in freshwater magnetotactic bacteria. *Science* **203**, 1355–1356.

Golubic, S. (1983). Stromatolites, fossil and recent: A case history. In *"Biomineralization and Biological Metal Accumulation"* (P. Westbroek and E. W. de Jong eds.), pp. 313–326. Riedel, Dordrecht, The Netherlands.

Greenfield, L. J. (1963). Metabolism and concentration of calcium carbonate by a marine bacterium. *Ann. N.Y. Acad. Sci.* **109**, 82–112.

Keefe, W., and Smith, J. (1977). Intracellular crystalline deposits by bacteria grown in urine from a stone former. *Invest. Urol.* **14**, 344–346.

Krumbein, W. E. (1979). Calcification by bacteria and algae. In *"Biogeochemical Cycling of Mineral-Forming Elements"* (P. A. Trudinger and D. J. Swaine, eds.), pp. 47–68. Elsevier, Amsterdam.

Margulis, L., and Stolz, J. F. (1983). Microbial systematics and a Gaian view of the sediments. In "Biomineralization and Biological Metal Accumulation" (P. Westbroek and E. W. de Jong, eds.), pp. 27–53. Riedel, Dordrecht, The Netherlands.

Nealson, K. H. (1983). Microbial oxidation and reduction of manganese and iron. In "Biomineralization and Biological Metal Accumulation" (P. Westbroek and E. W. de Jong, eds.), pp. 459–479. Riedel, Dordrecht, The Netherlands.

Streckfuss, J. L., Smith, W. N., Brown, L. R., and Campbell, M. M. (1974). Calcification of selected strains of *Streptococcus mutans* and *Streptococcus sanguis. J. Bacteriol.* **120**, 502–506.

Takazoe, I., Kurahashi, Y., and Turkuma, S. (1963). Electron microscopy of intracellular mineralization of oral filamentous microorganisms *in vitro. J. Dent. Res.* **42**, 681.

Towe, K., and Moench, T. T. (1981). Electron-optical characterization of bacterial magnetite. *Earth Planet. Sci. Lett.* **52**, 213–220.

Trudinger, P. A. (1979). The biological sulphur cycle. In "Biogeochemical Cycling of Mineral-Forming Elements" (P. A. Trudinger and D. J. Swaine, eds.), pp. 293–313. Elsevier, Amsterdam.

4

Eukaryotic Cells and the Accumulation of Ions

As we have seen, the prokaryotes had evolved lipid membranes, a complex organic metabolism, and a sophisticated information system based on nucleic acids. Their interactions with the environment, both singly and in ecological groupings, often produced mineral deposits. Within certain bacteria there were clearly the control systems and metabolism necessary to produce the mineralogically exact magnetosomes. They remained, however, morphologically simple and it was the evolution of a variety of compartments within the organism that released a more extensive biomineralization process.

I. Compartmentalization

The lipid membranes within which most biological processes occur produce their effects by acting as interfaces between different fluids. It was this membrane separation of volumes of fluids by the compartmentalization of intracellular and extracellular spaces that was a crucial development in biomineralization. Into these isolated spaces the organism was able to direct ions, control the composition of the fluid, and exceed the solubility product of minerals. If the volumes were small, the process could be achieved at relatively low energy costs.

Early in the development of the eukaryotic cell, several biochemical features became of such importance that they dictated the general nature of subsequent evolution. The first influence was an osmotic one related to the ionic composition of the cell. The *products* of the diffusible ions on two sides of a permeable membrane must always be equal if the system is to be in chemical equilibrium (Donnan effect). Unfortunately, this equilibrium is osmotically unstable since osmotic forces depend on the *sum* of the ions on each side of a membrane. This instability was overcome by the extensive extrusion of sodium ions from the cells, as in animals, or by resisting the tendency to burst by strengthening the cell walls with cellulose, as in plants. Clearly, the elaboration of ion pumps in the lipid membranes of the cell became increasingly important as cells became specialized and organelles evolved. A second feature of biochemical evolution was the use of phosphate compounds in energy metabolism. The most conspicuous example of this is, of course, adenosine triphosphate, which provides a universal energy source. In the process, a large number of phosphorylated intermediates of metabolism are produced which, being highly hydrophilic, are easy to retain within the lipid membrane-bound compartments of the cell. This provides a fundamental mechanism for controlling biochemical reactions. Since, however, calcium forms insoluble precipitates with phosphates, its presence is very disruptive and early in evolution it appears to have been expelled from the cell. Characteristically, the ionic calcium level of eukaryotic cells is around 10^{-8} mol dm^{-3}.

II. Ion Pumps

The movement or translocation of ions across cell membranes, often against electrochemical gradients, is so well known that biologists use such schemes in many of their conceptual models (Carafoli and Scarpa, 1982). Most theories of biomineralization invoke, often as an act of faith, a calcium pump, a proton pump, a source of bicarbonate or phosphate ions, or a silicic acid pump. Clearly, all of these activities in isolation or combination could account for processes that exceed the solubility product of a mineral. What, then, is the basis of these ion movements in eukaryotic cells?

The ion-motive ATPase enzymes discovered so far can be classified into three main types: F, P, and V (Fig. 4.1). Type F is involved in the synthesis of ATP from proton gradients of the sort that bacteria, chloroplasts, and mitochondrial membranes produce. As such, it is essential to all cells for the formation of ATP but it probably does not play any direct and intimate role in biomineralization. Type P ATPases are so-called

Figure 4.1 Types of ion-motive ATPases and their inhibitors (from Pedersen & Carafoli, 1987a).

P TYPE

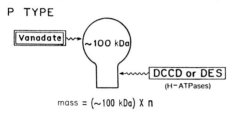

mass = $(\sim100$ kDa$)$ X n

V TYPE

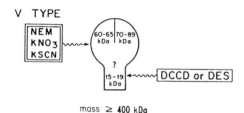

mass $\geq$ 400 kDa

F TYPE

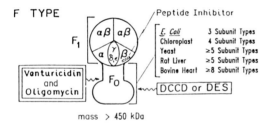

mass > 450 kDa

because they form a covalent phosphorylated intermediate on an aspartate residue of the enzyme as part of their reaction cycle. Examples are the Na^+/K^+-, Ca^{2+}-, and H^+- transporting enzymes of the plasma membranes of eukaryotic cells and the Ca^{2+} pump on the endoplasmic reticulum. All of these ATPases consist of a 70- to 100-kDa peptide that contains the phosphorylation and ATP-binding sites. The amino acid sequences of several of these enzymes have now been worked out and they show significant homologies, suggesting perhaps that there was a common ancestral molecule for Na^+/K^+- and Ca^{2+}-ATPases (Pedersen and Carafoli, 1987a). These enzymes have a cytoplasmic surface connected to a membrane-bound segment. There are high-affinity ion-binding sites in the cytoplasmic surface and it is thought that movement of

Figure 4.2   Relationships between the different types of ion-transporting ATPases in the cell showing how type F is involved in ATP synthesis and types P and V use ATP to translocate ions (see Fig. 4.1). (From Pedersen and Carafoli, 1987a.)

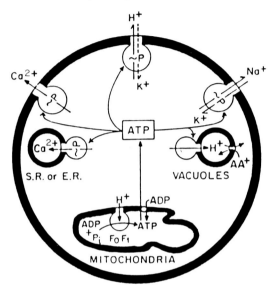

the protein helices occurs on phosphorylating the molecule (Pedersen and Carafoli, 1987b). This results in the formation of a channel and the extrusion of the ion onto the luminal surface (Fig. 4.2). Type V ATPases are mainly associated with vesicles such as the vacuoles of lower eukaryotes, the tonoplasts of plants, and the endosomes, clathrin-coated vesicles, and secretory granules of animal cells. These all appear to be proton-transporting enzymes that are made of several subunits with a total size of more than 400 kDa.

Despite a number of similarities between these ion-transporting ATPases, there are a number of distinctive properties as well as their ion specificities which allow them to be separated. The Na^+/K^+-ATPase is inhibited by ouabain and plasma membrane H^+-ATPase by dicyclohexylcarbodiimide (DCCD) and diethylstilbestrol (DES). The type P ATPases are blocked by vanadate while type V ATPases are more sensitive to tributyl tin, KNO_3, and KSCN (Pedersen and Carafoli, 1987a). It should be possible, therefore, to obtain experimental evidence for the role of these transport ATPases in biomineralization. Two examples will illustrate the types of experiment that are possible.

The silicification of the diatom *Navicula pelliculosa* has been known for a long time to require an energy component for the uptake of $Si(OH)_4$ (Lewin, 1955). Silicic acid is apparently transported on a carrier-mediated membrane system showing saturation kinetics with a dependence on cell metabolism (Sullivan, 1976). This is not, however, direct evidence for a silicification pump per se and more recent evidence shows a dependence on Na^+/K^+-ATPase. The inward flow of sodium ions apparently cotransports silica into the diatom forming what is referred to as a symport system (Bhattacharyya and Volcani, 1980). Thus, there is clear evidence for the active transport of ions in this mineralization process but it requires careful study to identify which of the ions are actively involved.

A second example of the importance of not assuming that a transport mechanism exists without first analyzing the system carefully is apparent in plants. From what we have said about the very low concentration of free Ca in the cytosol, it is immediately evident that the solubility product of the common minerals that are found within plant vacuoles will not be exceeded by simple diffusion from the cytosol. A calcium pump is clearly required in the vacuole wall but the exact nature of this pump is again uncertain. Two possibilities have been suggested (Macklon, 1984). One is a direct ATPase system, while the other is an antiport system (Fig. 4.3). The energy source for both mechanisms is the hydrolysis of ATP. In an antiport system, however, the *immediate* source of energy is an electrical or electrochemical gradient. Thus, Ca^{2+} transport

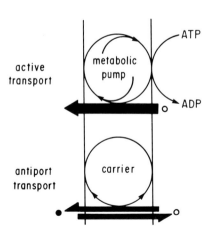

Figure 4.3 Two possible mechanisms for moving ions across the plasma membrane. In active transport (top) ATP is used as an energy source to move an ion against possible electrochemical gradients. In an antiport system (bottom) a carrier molecule facilitates the movement of an ion by exchanging with a counterion moving in the opposite direction.

against a concentration gradient could be accomplished by a gradient of protons which exchange across a membrane for Ca^{2+} moving in the opposite direction. A Ca^{2+}/H^+ antiport of this kind and a $Ca^{2+}/ATPase$ have both been identified in pea internode cells (Rasi-Caldogno et al., 1982) and in corn roots (Zocchi and Hanson, 1983). Vesicles of the plasmalemma of *Neurospora crassa* have also been shown to transport Ca by a Ca^{2+}/H^+ antiport in which the gradient of protons results from an electrogenic H^+-ATPase (Scarborough, 1982). In the tonoplast, the nature of the Ca transport mechanism, whether an antiport or a Ca^{2+}-ATPase or another, will only be resolved by isolating tonoplast membranes from mineralizing and nonmineralizing cells in sufficient quantity and purity for careful experimental studies (see Matile, 1978; Marty, 1982).

These examples will illustrate how the evolution of cellular compartments and the transport of cations across these membranes provide the cellular machinery that facilitates biomineralization. The basic mechanisms that are involved are fundamental to all forms of life and for that reason necessitate particular ingenuity by experimentalists who are trying to investigate the mechanisms of biomineralization. This should not, however, be used as an excuse for simply assuming that a specific cation pump exists wherever a mineral is deposited. As we have already seen, there could be a variety of systems that transport cations or, alternatively, it may be the supply of anions which leads to the onset of mineralization.

III. Anion Formation

Like cations, anions are involved in a wide range of cellular activities and the main skeletal minerals are, in fact, the salts of the principal inorganic buffers of the eukaryotic cell, i.e., the phosphate and bicarbonate systems. We have already seen the importance of proton gradients in synthesizing ATP and most cells regulate cytoplasmic pH within very close limits. Carbon dioxide, which is produced in large quantities during aerobic metabolism, is a small uncharged molecule which passes across cell membranes relatively easily. It reacts rather slowly with water to form carbonic acid that, being more polar, tends to become trapped at membrane barriers. The fluxes of carbon dioxide are influenced by the enzyme carbonic anhydrase that catalyzes the hydration and dehydration of carbon dioxide and occurs at numerous cellular sites in eukaryotes. It has a molecular mass of about 30 kDa and a size of roughly $4 \times 4.5 \times 5.5$ nm. There is one zinc atom per enzyme molecule. This zinc lies near the bottom of a 1.2-nm cleft in the surface of the enzyme and is bounded by the three imidazole groups of adjacent histidine molecules. The basic problem with the reaction

$$CO_2 + H_2O \rightarrow H_2CO_3 \tag{4.1}$$

is that at normal physiological pH values carbonic acid hardly exists. Even the reaction

$$CO_2 + OH^- \rightarrow HCO_3^- \tag{4.2}$$

is limited by the availability of hydroxy ions and is really only effective above pH 10. Clearly, in order to catalyze this reaction the enzyme needs a reactive group capable of inducing these reactions at pH 7 and this, in effect, is what the zinc atom does. By acting as a Lewis acid the metal attracts an electron pair from water, causing it to become polarized. The imidazole groups of the histidine bind onto the zinc, leaving one coordination site for it to react with the water (Fig. 4.4). As a result of this Lewis acid effect, the water is destabilized and loses a proton. This facilitates the attack on carbon dioxide and this catalytic effect is favored by the presence of a hydrophobic cleft. As a result, a $Zn-HCO_3$ complex is formed and subsequently displaced by a second water molecule (Coleman, 1984).

$$Zn + H_2O \xrightarrow{\uparrow H^+} + ZnOH^- + CO_2 \rightarrow ZnHCO_3 + H_2O \rightarrow Zn \cdot H_2O + HCO_3^- \tag{4.3}$$

and the overall effect is

$$H_2O + CO_2 \rightarrow H^+ + HCO_3^- \tag{4.4}$$

Aerobic metabolism produces carbon dioxide and drives the reactions in Eq. (4.4) to the right, releasing protons. It is therefore possible to determine the pH of the cell from its effect on the buffer carbonic acid. The dissociation constant (K_a) is defined as

$$K_a = a_{H^+} \times \frac{[salt]}{[acid]} \tag{4.5}$$

where a_{H^+} is the activity of protons, and which, if rearranged becomes

$$pH = pK_a + \log\frac{[HCO_3]}{[H_2CO_3]} \tag{4.6}$$

so relating pH to the ratio of bicarbonate and carbonic acid.

Photosynthesis, which removes carbon dioxide, drives the reactions of Eq. (4.4) to the left and tends to remove protons. It is now possible to

Figure 4.4 Structure of carbonic anhydrase showing the position of the zinc at the base of a cleft in the surface (arrowed). A possible interpretation of the role of the zinc is shown.

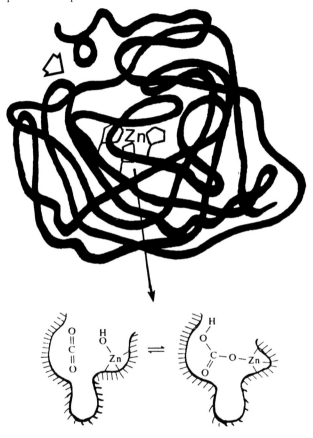

see how calcification is an elaboration on this system since, at physiological pH values, bicarbonate is the main anion, i.e.,

$$Ca + HCO_3^- \rightarrow CaCO_3 + H^+ \tag{4.7}$$

Sites of calcification are, therefore, acidotic and the process shown in reaction (4.7) will be affected by activities such as carbon dioxide production, the availability of bicarbonate, pH, and, of course, the enzyme carbonic anhydrase which catalyzes these interactions. Anything removing protons will, therefore, favor the precipitation of carbonates.

Biomineralization involving the deposition of phosphates will have the same effect. Alkaline phosphatases and various ATPases can all release phosphate ions and they are frequently described as being present at sites of mineralization. These enzymes could, therefore, be responsible for the local release of orthophosphates. At physiological pH ranges, the most prevalent form is HPO_4^{2-} and during calcification this is, in effect, converted to a calcium salt.

$$3Ca^{2+} + 2HPO_4^{2-} \rightarrow Ca_3(PO_4)_2 + 2H^+ \tag{4.8}$$

Protons will, therefore, be released and for calcification to continue they must, of course, be removed.

There are, therefore, close interactions between calcification, acid base metabolism, and a variety of enzymes. Any process that influences these activities could clearly interact through these systems.

The relationship between biomineralization and the cellular metabolism of anions is, therefore, both fundamental and complex. The presence of enzymes such as carbonic anhydrase or alkaline phosphatase may imply that they are acting as the necessary driving forces for the calcification process, but they both catalyze so many fundamental cellular processes that this may not be the case. Clearly, careful experimental evidence is again necessary before such enzymes are invoked as the basis for mineral deposition.

We can conclude, therefore, that biomineralization of the eukaryotes involves the movement of ions into isolated compartments of the organism. Some of these ion movements may be energy dependent, such as those involving transport ATPases, while others may be simply dependent on increasing the rate of formation of certain molecules such as the catalytic production of bicarbonate ions by carbonic anhydrase. In all cases, biomineralization involves the association of a few fundamental properties of cell biology which have been combined in a variety of ways in different species. It is, however, up to the experimentalist to demonstrate that there is a causal relationship between these cellular processes and the phenomenon of biomineralization. Such correlations should not be invoked *a posteriori* but should be used as opportunities for investigating the fundamental mechanisms that exist.

References

Bhattacharyya, P., and Volcani, B. E. (1980). Sodium-dependent silicate transport in the apochlorotic marine diatom, *Nitzschia alba. Proc. Natl. Acad. Sci. U.S.A.* **77,** 6386–6390.

Carafoli, E., and Scarpa, A. (eds.) (1982). Transport ATPases. *Ann. N.Y. Acad. Sci.* **402**, 1–604.

Coleman, J. E. (1984). Carbonic anhydrase: Zinc and the mechanism of catalysis. *Ann. N.Y. Acad. Sci.* **429**, 26–48.

Lewin, J. C. (1955). Silicon metabolism in diatoms. III. Respiration and silicon uptake in *Navicula pelliculosa*. *J. Gen. Physiol.* **39**, 1–10.

Macklon, A. E. S. (1984). Calcium fluxes at protoplasmalemma and tonoplast: A review. *Plant Cell Environ.* **7**, 407–413.

Marty, F. (1982). Isolation, freeze-fracture and characterization of vacuole membrane fragments. *In* "Plasmalemma and Tonoplast: Their Functions in the Plant Cell" (D. Marme, E. Marme, and R. Hertel, eds.), pp. 179–188. Elsevier, New York.

Matile, P. (1978). Biochemistry and function of vacuoles. *Ann. Rev. Plant Physiol.* **29**, 193–213.

Pedersen, P. L., and Carafoli, E. (1987a). Ion motive ATPases. I. Ubiquity, properties and significance to cell function. *Trends Biochem Sci.* **12**, 146–150.

Pedersen, P. L., and Carafoli, E. (1987b). Ion motive ATPases. II. Energy coupling and work output. *Trends Biochem. Sci.* **12**, 186–189.

Rasi-Caldogno, F., de Michelis, M. I., and Pugliarello, M. C. (1982). Active transport of Ca^{2+} in membrane vesicles from pea. *Biochim. Biophys. Acta* **693**, 287–295.

Scarborough, G. A. (1982). The *Neurospora* plasma membrane ATPase. *In* "Plasmalemma and Tonoplast: Their Functions in the Plant Cell" (D. Marme, E. Marme, and R. Hertel, eds.), pp. 431–438. Elsevier, New York.

Sullivan, C. W. (1976). Diatom mineralization of silicic acid. I. $Si(OH)_4$ transport characteristics in *Navicula pelliculosa*. *J. Phycol.* **12**, 390–396.

Zocchi, G., and Hanson, J. B. (1983). Ca transport and ATPase activity in a microsomal vesicle fraction from corn roots. *Plant Cell Environ.* **6**, 203–209.

5

The Control of Mineralization

Paradoxically, the most conspicuous aspect of biomineralization is the feature which biologists have tended to ignore most. The crystals that chemists and geologists study are symmetrical structures with precise angles between flat crystal faces. Biomineralization, on the other hand, produces delicate fenestrated baskets, curved teeth, irregular statoliths, and sculptured shells formed of myriad crystals in various arrangements. Somehow the geometrical shapes of crystallography are modified to form the functional structures of biology. It is easier to see how this could be achieved with amorphous minerals. As their name suggests, they are freed from the constraints of a geometrical repeating unit and this may be one of the reasons silica is used by so many of the lower plants and animals to form the elaborate tests and frustules. With crystals, however, the system is more complex, for it implies control over the processes of solid deposition that may very well lie at the heart of the process of biomineralization. It is likely that this control involves both organic and inorganic molecules and that both kinetic and thermodynamic influences are implicated.

I. Organic Matrix

The organic matrix is usually considered to be the material intimately associated with and present within mineralized structures. It is, therefore, distinct from the sclerotized outer coverings of invertebrates such as the periostracum and cuticle, although these may have a close relationship to the crystals of exoskeletons. Usually, the term matrix is used without necessarily implying any particular function. However, one very broad definition suggests that a matrix is any organized surface that acts as a mediator of mineralization (Mann, 1983).

The present discussion of matrix will be general rather than specific since we shall have to return to matrices in considering individual taxonomic groups. From the start it is important to mention that the organic material within mineral structures may be complex and is not of uniform composition in different taxa. Protein, glycoprotein, carbohydrates such as glycosaminoglycans, and lipid have been shown to be present in a variety of mineralized matrices (Wilbur and Simkiss, 1968; Wilbur and Manyak, 1984). It is worth emphasizing, however, that no complete analysis has yet been carried out on a single species. Accordingly, no matrix can be precisely defined in terms of its chemistry. Within marine algae, carbohydrate is the principal component (de Jong, 1975). In invertebrate exoskeletons, two major organic fractions can be separated; a water-soluble and a water-insoluble fraction (Weiner and Lowenstam, 1983; Kingsley and Watabe, 1984). The soluble fraction of various species of molluscs has been extensively analyzed and found to contain many fractions with end groups that bind calcium (see Wilbur and Manyak, 1984). Inevitably, the isolation and separation procedures that are necessary to extract the organic matrix will disrupt the molecular organization that exists *in situ* and that may be pertinent to many mineral–protein interactions. It is possible, however, by means of X-ray and electron diffraction studies of molluscan matrix, to show well-defined relationships in the orientations of chitin, insoluble protein, and inorganic crystals (Weiner and Traub, 1981, 1984).

The organic matrix has been implicated as the causal agent for a large number of the characteristic features of biomineralization. Since these recur in many discussions of biomineralization, it is convenient to list them here.

1. Anionic groups on the matrix organic matter could concentrate calcium ions at various sites and thereby induce the supersaturation necessary for mineral nucleation. There is evidence that cations are bound as loose chelates (Tyler and Simkiss, 1958), producing an effect that is similar to a fixed charge in a Donnan equilibrium. This concentration of inorganic ions on the surface of the matrix molecule is the basis of the ionotropy theory for mineral deposition which is

gaining increased support (Crenshaw and Ristedt, 1976; Greenfield *et al.*, 1984) (Fig. 5.1).

2. Soluble matrix proteins inhibit mineral deposition (Wheeler *et al.*, 1981). This has been demonstrated in a number of experimental systems (Borman *et al.*, 1983) and is apparently due to the protein binding onto the crystal lattice. Such molecules could, therefore, be used to control the process of mineralization.

3. Matrix proteins can favor the growth of particular isomorphs or inhibit the growth of certain crystal faces (Addadi and Weiner, 1985). In many ways this is a more specific form of (2) since it involves the protein having a steric configuration whereby it binds to a particular crystal or inhibits the growth of a specific face of the crystal lattice. It could, therefore, control the types of crystals that are formed in mineralization as has recently been shown by compressing membrane monolayers (Mann *et al.*, 1988).

4. The soluble matrix proteins may become overgrown by mineral and trapped in the crystal. This would influence the strength of the crystal (Silyn-Roberts and Sharp, 1986) and the development of microcrystalline structures around the inclusion. Such domains may enable the mineral to be shaped more easily into the structures seen in biological skeletons (Simkiss, 1986).

5. Insoluble matrix may form a structural framework which is subsequently covered by a layer of the more reactive soluble matrix (Fig. 5.2). This "sandwich" theory of soluble and insoluble matrix components was originally suggested by Degens (1978) and has subse-

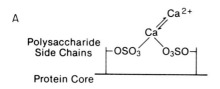

Figure 5.1 Ionotropic binding. (A) High affinity calcium binding is indicated. (B) Carbonate and secondary calcium binding occurs (Greenfield *et al.*, 1984).

Figure 5.2 The "epitactic matrix" theory of mineralization based on the presence of an "insoluble protein" which forms the structural frame covered by a "soluble protein" capable of nucleating inorganic minerals. The top illustration shows how aspartate residues could correspond to an aragonite lattice while the bottom illustration suggests a similar interaction between silicic acid and serine residue(s) in a polypeptide chain. (After Simkiss, 1986.)

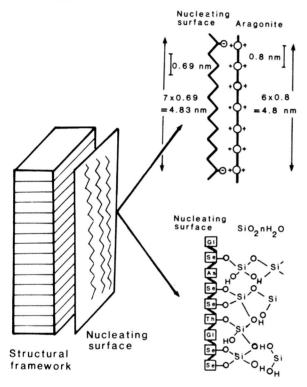

quently been supported by ultrastructural and biochemical studies (Weiner, 1986).

Matrix deposition precedes or accompanies crystal formation and often occupies the medium in which crystals form and grow. Each of these situations is, of course, capable of having a different influence on the mechanism of crystal growth. In several invertebrate phyla the matrix is secreted by epithelial cells contiguous to the site of mineral deposition in the exoskeleton. In the case of mineral deposition in plant and animal cell vacuoles a different situation exists. The organic molecules that form the matrix are contained in small vesicles in the cytoplasm and are passed into the mineralizing vacuoles by the fusion of their membranes.

A number of invertebrate exoskeletons are constructed of layers of crystals and examination of these structures clearly shows that matrix material is present between the crystal layers and envelopes individual crystals. By means of a diamond knife and considerable perseverance, individual crystals can also be sectioned. Demineralization of these sections reveals that matrix material can be incorporated within crystals during their growth (Watabe, 1965; Dunkelberger and Watabe, 1974), which implies that the true unit of biomineralization may consist of these individual microcrystallites (Parker *et al.*, 1983). The incorporation of the organic material within the crystal structure clearly results from crystal formation and growth within a medium that contains the matrix molecules.

Four functions have therefore been attributed to organic matrix: (1) crystal nucleation, (2) crystal orientation, (3) crystal morphology including size, and (4) control of polymorphic type. The first three of these possible functions will be considered in the following sections.

II. Organic Matrix and Nucleation

Nucleation will be facilitated by the formation of bonds between mineral ions and organic matrix molecules. With the formation of bonds, the hydration layer of the ions will in part be removed. A consequence of this interaction between mineral ions and matrix is a decrease in the activation energy of nucleation (Mann, 1983; Williams, 1984), which is the free energy barrier that must be surmounted for nucleation to occur.

Certain matrix proteins of molluscan shell are thought to initiate nucleation because of their calcium-binding capacity. The binding has been explained as due to (1) a high content of aspartic and glutamic acids with COO^- groups (Weiner, 1979; Weiner and Traub, 1981), (2) the ester sulfate groups which provide a negative charge on hexosamine residues (Crenshaw, 1982), or (3) a specific amino acid sequence (Samata *et al.*, 1980). The bound Ca ions are presumed to attract CO_3^{2-} ions and, by having a sufficient concentration of these ions, to induce nuclei formation. In this way the nuclei would be fixed and oriented on the protein substrate. Addadi and Weiner (1985) have observed that a purified acid-rich protein from molluscan shell matrix, when adsorbed onto an artificial substrate, would nucleate crystals on its exposed surface with a specific crystal face fixed to the protein. Unfortunately, the specificity of this "isotropic" growth may not be very great but the mechanism is of considerable interest. Other evidence that the organic matrix can initiate crystal formation has been obtained by completely decalcifying skeletal material and exposing it to saline solutions containing calcium. Matrix of molluscan shell, the organic material from calcareous tubes of serpulids, and gorgonian spicules all exhibited recalcification in this situation (Watabe and Wilbur, 1960; Wilbur and Watabe, 1963; Bernhardt *et al.*, 1985; Watabe *et al.*, 1986). In such *in vitro* tests of recalcification, it is to be expected that the matrix will be different from that *in situ* in that soluble portions may have been removed in the decalcification procedure and the ionic nature of the matrix may then be altered. At best therefore these experiments are only suggestive of the influence that matrix materials can have *in situ*.

III. Organic Matrix and Crystal Orientation

A vertical section through many invertebrate skeletons such as molluscs and certain brachiopods shows an ordered arrangement of small crystals in layers, each layer one crystal in thickness. The crystals in any one region usually have a similar orientation of one or more crystallographic axes. Here, then, are two characteristics of mineral deposition: layers of small crystals and uniformity of orientation of crystallographic axes. The counterpart in plants is an ordered arrangement of crystals within the cell vacuole, although it should be stressed that in neither invertebrates nor plants is crystal ordering universal. Thus, for example, instead of uniformly ordered crystals, aggregates of spherulites are found within parts of the skeleton of corals, bryozoans, molluscs, sipunculids, and brachiopods (Watabe, 1981; Pan, 1985).

The uniformity of orientation of crystallographic axes in exoskeletons has usually been attributed to the influence of the molecules of the organic matrix. This is a reasonable assumption in that crystal growth takes place in close association with matrix molecules. The X-ray and electron diffraction studies previously mentioned show that crystal orientation in molluscan shell bears a clearly defined relation to the orientation of the insoluble matrix molecules and thus supports this concept (Weiner and Traub, 1981, 1984).

Phenomena of this type are sometimes considered as evidence for epitaxy. In the strict sense, epitaxy refers to oriented crystal growth in which one crystal lattice overgrows another with similar dimensions. This orientation may begin with an inorganic crystal nucleus formed directly on oriented organic molecules or it may occur through later contact with organic molecules not involved in nucleation. Epitaxy that has its origin at the time of nucleation will take place when the activation energy for nucleation is minimal for one of the crystal faces (Mann, 1983). If subsequent growth of the crystal nuclei is to be oriented rather than nonoriented, a high level of supersaturation may be required.

Oriented crystal growth need not require a high degree of matching of the crystal lattice and substrate molecules. Different polymorphs of calcium carbonate can be induced to form on surfaces of stearic acid by partially compressing the film but the results suggest that some mobility of the charges on the organic surface may be beneficial (Mann *et al.*, 1988). But whether the growth is epitaxial or the lattice and substrate are less precisely matched, the subsequent orientation of the crystal axes will be governed by the direction in which the energy of growth is at a minimum. The actual region of contact between a crystal and a substrate may, through distortion, tolerate a mismatch such that the orientation of the crystal is not affected. In fact, if the mismatch is great enough and if there are many dislocations, it may again bring regions of the crystal lattice into register with the substrate unit cell spacings.

In considering crystal growth and its orientation, one is often inclined to view crystals as growing on a planar substrate in solution. Such would be true, for example, of crystals growing on previously deposited crystals or on a surface sclerotized by enzyme action. More commonly, however, crystals grow *within* a matrix medium secreted by cells. In doing so, the orientation of the crystals, whether epitaxial or nonepitaxial, may well be determined by elongate matrix molecules. Consequently, as the crystals grow three-dimensionally, the matrix molecules will be displaced.

Now imagine the following hypothetical situation. Many nucleation centers develop within the matrix that occupies a microspace. As a result of this matrix influence, the lattice of each nucleus becomes similarly oriented and the nuclei grow into crystallites. Assume, then, that the growing crystallites fill the space originally occupied by the matrix and in the process compress the matrix molecules between neighboring crystallites so as to make them less apparent. This mosaic of crystallites, all with the same orientation, may have certain properties of a single crystal and be so interpreted. But whether a skeletal unit is a single crystal or a polycrystalline aggregate may be difficult to determine experimentally. It seems likely, therefore, that the mosaic structure of some skeletal units explains the controversy as to the exact crystalline nature of the echinoderm skeleton and a similar situation seems to exist with otoconia from the inner ear. The otoconia were originally described as single crystals (Carlstrom, 1963) but are now known to have a mosaic structure (Mann *et al.*, 1983). However, the discussion has not been a sterile one, since it is easy to envisage how the presence of crystalline domains enables a skeleton to be shaped into a number of diverse structures. Unlike large geometrical crystals, these microdomains are more easily molded into a variety of biological shapes. Similar results may, however, also be produced by other influences.

IV. Inorganic Modifiers of Crystal Growth

Since the growth of a crystal depends on the rate at which it is supplied with ions, its shape can easily be modified by a whole range of influences (Nielsen and Christoffersen, 1982). There are at least five stages in the biomineralization process where this could be affected, namely:

1. The rate of supply of ions to the crystallizing solution may be modified by influencing the movement of ions around or through cell membranes.
2. The diffusion of ions to the growing crystal surface may be controlled. The distance (l) that is moved in time t depends on the diffusion coefficient D, i.e.,

$$l^2 = 2Dt$$

   The value for the diffusion coefficient of calcium in the cytoplasm is roughly $10^{-7} cm^2 sec^{-1}$ from which it can be calculated that the ion would pass across a cell within a matter of about 1 sec and through a solution considerably faster. Since the formation of biominerals takes place within microspaces the path length and hence the time involved will be very small. Thus the rate of diffusion in the immediate vicinity of the crystal face will govern crystal growth and variations in the rate of diffusion in the region around the crystal would greatly influence the shape of the mineral that is formed.
3. *Adsorption:* This refers to the incorporation of ions onto the growing crystal surfaces. The crystal surface will be bombarded by free ions in the crystallizing medium, and there will also be an attraction and repulsion of these ions by the charges of the ions in the crystal lattice. Since the free ions are hydrated, at least some of the water of hydra-

tion must be removed for the ions to approach the crystal surface closely enough to become incorporated into the lattice. To remove the water of hydration from these ions, an activation energy is required. If these energy barriers are not immediately overcome, a highly hydrated mineral may form that will slowly transform into a more stable crystal.

4. *Integration:* Ions become added to the crystal lattice and increase crystal size at growth sites of the crystal. These sites are primarily the steps, kinks in the steps, and screw dislocations mentioned on pp. 17–19. The rate of adsorption onto the crystal will increase with the number of these sites.

5. *Inhibitors:* The presence of inhibitors may slow the rate of nucleation and inhibit the addition of ions into the crystal lattice. The inhibitory substances may be inorganic molecules, including phosphates, organic molecules present in the cells and fluids of organisms (see, for example, Table 5.1), and foreign substances of various kinds, including heavy metals. Inhibitors may react with or cover the lattice and so interfere with the adsorption of other ions onto the lattice. But since many inhibitors are effective at extremely low dilutions, it is assumed that they often work by blocking the specific sites of growth at kinks. Compounds that inhibit the formation of crystals of calcium carbonate and hydroxyapatite are listed in Table 5.1.

In addition to these chemical effects, various rigid and nonrigid surfaces may limit crystal growth. The crystals that are found in plant vacuoles have often been described as developing within preformed chambers of organic material (p. 46). These chamber walls could restrict

Table 5.1 Biological Inhibitors of Hydroxyapatite Formation in Aqueous Solution[a]

Mg^{2+}
CO_3^{2-}
Pyrophosphate ($P_2O_7^{4-}$)
Polyphosphates: EHDP, Cl_2MDP
Nucleotide polyphosphates: adenosine triphosphate, guanosine diphosphate, glucose 1,6-diphosphate
Cartilage proteoglycans
Dentine phosphoproteins
Polycarboxylates ($RCO_2)_n$: polyglutamate, polyacrylate
Phospholipids
3-Phosphocitrate: $[(CH_2CO_2)C(CO_2)OPO_3]^{4-}$

[a] From Mann (1983).

growth and control crystal size by interfering with additions to the lattice as it comes in contact with them. A similar mechanism of inhibition may perhaps occur in intravacuolar mineralization in animal cells. However, vacuole membranes may also enlarge as the crystals grow and contact them. The cytoskeleton, by pulling on the vacuole membrane, may thus shape the mineral that is formed within it. In an analogous way sclerotization of the secreted matrix present between crystal layers in molluscan shell could also provide a barrier to crystal growth and thus determine the structure of the shell.

Biocrystals of a single mineral type may exist in a great range of forms as a result of these modifications to crystal growth. Coprecipitation of ions, modifications to lattice dimensions, and the control over the growth of particular crystal faces can all be used to influence the shape of mineral deposits. These effects are undoubtedly used by organisms to produce the elaborate skeletons that are found in nature. There are, however, other influences that also appear to leave an effect on mineral deposition. Temperature can have a considerable influence on the type of crystal polymorph that is formed. The ratios of the polymorphs of $CaCO_3$ that are deposited in coccoliths of the marine alga *Emiliania huxleyi* depend fairly directly on the temperature at which they are cultured (Wilbur and Watabe, 1963). This phenomenon is of considerable interest to geologists since the form and composition of such biomineralized deposits may then be used to indicate environmental temperatures many millions of years ago (Lowenstam, 1954).

References

Addadi, L., and Weiner, S. (1985). Interactions between acidic proteins and crystals: Stereochemical requirements in biomineralization. *Proc. Natl. Acad. Sci. U.S.A.* **82,** 4110–4114.

Bernhardt, A. M., Kunigelis, S. C., and Wilbur, K. M. (1985). Effects of phosphates on shell growth and calcium carbonate crystal formation. *Aquat. Toxicol.* **7,** 1–13.

Borman, A. H., de Jong, E. W., Huizinga, M., and Westbroek, P. (1983). Inhibition of $CaCO_3$ precipitation by a polysaccharide associated with coccoliths of *Emiliania huxleyi*. In "Biomineralization and Biological Metal Accumulation" (P. Westbroek and E. W. de Jong, eds.), pp. 303–305. Riedel, Dordrecht, The Netherlands.

Carlstrom, D. (1983). A crystallographic study of vertebrate otoliths. *Biol. Bull. (Woods Hole, Mass.)* **110,** 159–169.

Crenshaw, M. A. (1982). Mechanisms of normal biological mineralization of calcium carbonates. In "Biological Mineralization and Demineralization" (G. H. Nancollas, ed.), pp. 243–257. Springer-Verlag, Berlin.

Crenshaw, M. A., and Ristedt, H. (1976). The histochemical localization of reactive groups in the septal nacre from *Nautilus pompilus*. *In* "The Mechanisms of Mineralization in the Invertebrates and Plants" (N. Watabe and K. M. Wilbur eds.), pp. 335–367. Univ. of South Carolina Press, Columbia, South Carolina.

Degens, E. T. (1976). Molecular mechanisms of calcium phosphate and silica deposition in the living cell. *Top. Curr. Chem.* **64**, 1–112.

de Jong, E. W. (1975). "Isolation and Characterization of Polysaccharides Associated with Coccoliths," Ph.D. thesis. State University of Leiden, Leiden, The Netherlands.

Dunkelberger, D. G., and Watabe, N. (1974). An ultrastructural study on spicule formation in the Pennatulid colony *Renilla reniformis*. *Tissue Cell* **6**, 573–586.

Greenfield, E. M., Wilson, D. C., and Crenshaw, M. A. (1984). Ionotropic nucleation of calcium carbonate by molluscan matrix. *Am. Zool.* **24**, 925–932.

Kingsley, R. J., and Watabe, N. (1984). Synthesis and transport of the organic matrix in the gorgonian *Leptogorgia virgulata* (Lamark) (Coelenterata: Gorgonacea). *Cell Tissue Res.* **235**, 533–538.

Lowenstam, H. A. (1954). Environmental relations of modification compositions of certain carbonate secreting marine invertebrates. *Proc. Natl. Acad. Sci. U.S.A.* **40**, 39–48.

Mann, S. (1983). Mineralization in biological systems. *Struct. Bonding (Berlin)* **54**, 125–174.

Mann, S., Heywood, Brigid R., Rajam, S. and Birchall, J. D. (1988). Controlled crystallization of $CaCO_3$ under stearic acid monolayers. *Nature (London)* **334**, 692–695.

Mann, S., Parker, S. B., Perry, C. C., Ross, M. D., Skarnulis, A. J., and Williams, R. J. P. (1983). Problems in the understanding of biominerals. *In* "Biomineralization and Biological Metal Accumulation" (P. Westbroek and E. W. de Jong, eds.), pp. 171–183. Riedel, Dordrecht, The Netherlands.

Nielsen, A. E., and Christoffersen, J. (1982). The mechanisms of crystal growth and dissolution. *In* "Biological Mineralization and Demineralization" (G. H. Nancollas, ed.), pp. 79–99. Springer-Verlag, Berlin.

Pan, C. M. (1985). "Ultrastructural and Physiologial Investigations of the Mineralization of the Inarticulate Brachiopod, *Glottidia pyramidata* (Stimpson)," Ph.D. thesis. University of South Carolina, Columbia, South Carolina.

Parker, S. B., Skarnulis, A. J., Westbroek, P., and Williams, R. J. P. (1983). The ultrastructure of coccoliths from the marine alga *Emiliania huxleyi* (Lohman). *Proc. R. Soc. London, Ser. B* **219**, 111–117.

Samata, T., Sanguansri, P., Cazaux, C., Hamm, M., Engels, J., and Krampitz, G. (1980). Biochemical studies on components of mollusc shells. *In* "The Mechanisms of Biomineralization in Animals and Plants (M. Omori and N. Watabe, eds.), pp. 37–47. Tokai Univ. Press, Tokyo, Japan.

Silyn-Robers, H., and Sharp, R. M. (1986). Crystal growth and the role of the organic network in eggshell biomineralization. *Proc. R. Soc. London, Ser. B.* **227**, 303–324.

Simkiss, K. (1986). The processes of biomineralization in lower plants and animals—An overview. *Syst. Assoc. Spec. Vol.* **30**, 19–37.

Tyler, C., and Simkiss, K. (1958). Reactions between eggshell organic matrix and metallic cations. *Q. J. Microsc. Sci.* **99**, 5–13.

Watabe, N. (1965). Studies on shell formation. XI. Crystal–matrix relationships in the inner layers of mollusk shells. *Ultrastruct. Res.* **12**, 351–370.

Watabe, N. (1981). Crystal growth of calcium carbonate in the invertebrates. *Prog. Cryst. Growth Charact.* **4**, 99–147.

Watabe, N., and Wilbur, K. M. (1960). Influence of the organic matrix on crystal type in molluscs. *Nature (London)* **188**, 334.

Watabe, N., Bernhardt, A. M., Kingsley, R. J., and Wilbur, K. M. (1986). Recalcification of decalcified spicule matrices of the gorgonian *Leptogorgia virgulata* (Cnidaria: Anthozoa). *Trans. Am. Microsc. Soc.* **105,** 311–318.

Weiner, S. (1979). Aspartic acid-rich proteins: Major components of the soluble organic matrix of mollusc shells. *Calcif. Tissue Int.* **29,** 163–167.

Weiner, S. (1986). Organization of extracellularly mineralized tissues: A comparative study of biological crystal growth. *CRC. Crit. Rev. Biochem.* **20,** 365–408.

Weiner, S., and Lowenstam, H. A. (1983). Organic matrix in calcified exoskeletons. *In* "Biomineralization and Biological Metal Accumulation" (P. Westbroek and E. W. de Jong, eds.), pp. 205–244. Riedel, Dordrecht, The Netherlands.

Weiner, S., and Traub, W. (1981). Organic–matrix–mineral relationships in mollusk shell nacreous layers. *In* "Structural Aspects of Recognition and Assembly in Biological Macromolecules" M. Balaban, J. L. Sussman, W. Traub, and A. Yonath, eds.), pp. 467–482. Balaban ISS, Rehovot, Israel.

Weiner, S., and Traub W. (1984). Macromolecules in mollusc shells and their functions in biomineralization. *Philos. Trans. R. Soc. London, Ser. B* **304,** 425–434.

Wheeler, A. P., George, J. W., and Evans, C. R. (1981). Control of calcium carbonate nucleation and crystal growth by soluble matrix of oyster shell. *Science* **212,** 1397–1398.

Wilbur, K. M., and Manyak, D. M. (1984). Biochemical aspects of molluscan shell mineralization. *In* "Marine Biodeterioration: An Interdisciplinary Study" (J. D. Costlow and R. C. Tipper, eds.), pp. 30–37. Naval Inst. Press, Annapolis, Maryland.

Wilbur, K. M., and Simkiss, K. (1968). Calcified shells. *Compr. Biochem.* **26A,** 229–295.

Wilbur, K. M., and Watabe, N. (1963). Experimental studies on calcification in molluscs and the alga *Coccolithus huxleyi. Ann. N.Y. Acad. Sci.* **109,** 82–112.

Williams, R. J. P. (1984). An introduction to biominerals and the role of organic molecules in their formation. *Philos. Trans. R. Soc. London, Ser. B.* **304,** 411–424.

PART 2

Cellular Organizations

6

Protoctista—Secreted Sculptures

The protoctista are eukaryotes, i.e., organisms with chromosomes and nuclei but without the elaboration into the multicellular structures that are characteristic of the main groups of plants and animals. They are often classified differently by botanists and zoologists but they include the protozoa, coccolithophorids, diatoms, radiolarians, and foraminifera. As such, they include 28 phyla (Margulis and Stolz, 1983) of which 19 have mineral-forming species. They are the main sediment-forming organisms of the oceans, producing minerals as different as barite $(BaSO_4)$, celestite $(SrSO_4)$, gypsum $(CaSO_4 \cdot 2H_2O)$, and whewellite (calcium oxalate) (Lowenstam, 1986). It is, however, as the main producers of silica and calcium carbonate tests (or exoskeletons) that they have attracted most attention. Remote sensing satellites frequently detect "blooms" of these organisms in the oceans of the world. One, which covered an area of 7200 km^2 was conservatively estimated to contain 70,000 tons of calcite in the upper layers of the sea (Holligan et al., 1983) and the extensive beds of cretaceous chalk and limestone suggest that such phenomena have been even more important in earlier ages.

It is, however, not only the mass of mineral produced by these organisms that is impressive. The mineralized structures that they deposit are among the most elaborate and delicate structures known; and, in some species, it is possible to trace their secretion from within the cells to their extrusion onto the surface where they may form an intricate interlocking exoskeleton. For this reason, they have become favored systems for the study of the physiology and biochemistry of biomineralization. Therefore, in this chapter, we shall initially give an indication of the diversity of mineral deposits produced by the protoctista and then concentrate on the wealth of experimental information on silicification and calcification.

I. Protozoa and the Diversity of Minerals

There are four main groups of protozoa: the Mastigophora (Zoomastigina), which move with flagellae and are thought to be the most primitive; the Sarcodina or amoeboid protozoa (including the Foraminifera, Radiolaria, and Rhizopoda); the ciliated Ciliophora; and the parasitic Sporozoa. Good skeletal structures are formed in most of these groups (Table 6.1) but there are intracellular granules in many others.

A. Calcium, Strontium, and Barium Minerals

In the ciliate *Tetrahymena*, which is easily cultured and, therefore, perhaps the best studied of the protozoa, there are large numbers of so-called volutin granules. These have long been regarded as storage granules but, in a detailed and careful piece of work, Rosenberg (1988) extracted them and showed that they consisted of calcium and magnesium pyrophosphates ($CaMgP_2O_7$). Subsequently, Nilsson and Coleman (1977) showed that these deposits accumulated a number of other trace metals and speculated that they might be involved in intracellular ion regulation. The system is, however, undoubtedly complex and the pyrophosphate granules are most common when growth is suboptimal or the organism is exposed to adverse nutritional conditions (Jones et al., 1984). The pyrophosphate granule recurs in a number of quite unrelated animals and it appears that these membrane-enclosed minerals are physiologically quite distinct from other skeletal deposits. A wide variety of calcium deposits were described in ciliates by Faure-Fremiet (1957) and one of the best-studied examples is *Spirostomum ambiguum*, a common organism in rivers and lakes. Specimens taken from old cultures of this animal often contain numerous granules of calcium phosphate (Jones, 1967) which, like those in *Tetrahymena*, tend to be lost when the protozoan is kept in rapidly dividing cultures. According to Pautard (1970), mineral deposition occurs when a number of small vesicles coalesce to produce a larger vacuole in which the granule is formed. The membrane of this vacuole may be important in initiating mineralization but a much better example of a well-developed calcified skeleton occurs in the ciliate *Coleps*. In this genus the mineral consists of a calcium phosphocarbonate lattice of longitudinal and transverse bars secreted into the alveolar spaces beneath the cell membrane (Faure-Fremiet *et al.*, 1968). The initial material is amorphous, and the calcium is deposited onto an organic matrix which is then shaped to form the final lattice.

It will be apparent from this brief introduction that protozoa are able to accumulate mineral deposits within their cytoplasm. What is not clear, however, is the extent to which these granules are skeletal deposits or even accumulations of ions as opposed to metabolic products of energy (pyrophosphate) or nutrient (phosphate) storage. There are, however, other kinds of mineral deposits in protozoa that are much easier to interpret.

A number of radiolarians produce flagellated isospores with a large vacuole that contains a crystal of strontium sulfate (Hollande and Martoja, 1974). The function of this crystal is not known but it may assist settling in the ocean during reproduction (Hanson, 1967). Since acantharian protozoa also possess a skeleton of strontium sulfate crystals, the presence of this mineral has been used to support the proposition that the two groups possess a common ancestor (Anderson, 1981). Unfortunately, little is known about the secretion of these strontium sulfate deposits. Thus, although the arrangement of the spicules is related to the crystallography of strontium sulfate the morphology of the skeleton is clearly under biological control (Perry *et al.*, 1983). The crystals of barium sulfate that are formed by the ciliate *Loxodes* are also poorly understood (Rieder *et al.*, 1982). It is difficult, however, to envisage an explanation for the formation of either of these deposits that does not involve a metabolic mechanism for transporting these ions through the cell and across the vesicle membrane.

Among the Sarcodina, the foraminifera are examples of some of the most elaborate mineralizers. These protozoa live within chambered shells, which they produce in such numbers that they have been responsible for many of the limestone geological strata. The process of shell formation has a number of interesting aspects. It is initiated by the formation of a "primary organic membrane" outside the cytoplasm. On the outer and inner surfaces of this are deposited additional organic layers to form a matrix or compartment within which the calcite layers are secreted (Anderson and Bé, 1978). The role of this organic layer and the way that it might act as a template for mineral deposition or as a compartment within which calcification occurs have been discussed in detail by Towe (1972).

A great variety of calcified structures are formed among the foramini-

Table 6.1 The Diversity, Distribution, and Localization of Biogenic Minerals in Extant Protoctista[a]

Mineral	Phylum	Lower taxa	Mineralization site[b]
Calcite (CaCO$_3$)	Myxomycota	*Didimium*	Ext
	Ciliophora	*Spirostomum*	Int
	Rhizopoda	*Paraquadrula*	Int
	Foraminifera	Carbonate superfamilies except Robertinacea	Ext
	Dinoflagellata	*Scrippsiella*	Ext
	Zoomastigina	*Pseudokephyrion*	Int
	Haptophyta	All genera with mineralized hard parts	Int
			Ext
	Charophyta	In most genera	Ext
	Rhodophyta	Genera in Cryptonemiales	Ext
Aragonite (CaCO$_3$)	Foraminifera	Genera of Robertinacea	Ext
	Rhodophyta	*Liagora, Galaxaura*	Ext
	Chlorophyta	Genera in Dasycladales, Caulerpales	Ext
	Phaeophyta	*Padina*	Ext
Vaterite (CaCO$_3$)	Rhodophyta	*Galaxaura*	Ext
"Carbonate"	Gamophyta	*Oocardium*	Ext
Dahllite [Ca$_5$(PO$_4$ CO$_3$)$_3$OH]	Ciliophora	*Spirostomum*	Int
ACP	Rhizopoda	*Cryptodifflugia*	?Int
Whewellite (CaC$_2$O$_4 \cdot$ H$_2$O)	Chlorophyta	*Penecillus, Udotea, Rhipocephalus*	Int
"Ca^{2+} oxalate"	Gamophyta	*Spirogyra*	Int
	Chlorophyta	*Acetabularia, Bornetella*	Int

Gypsum (CaSO$_4 \cdot$2H$_2$O)	Gamophyta	*Closterium, Penium, Pleurotaenium, Telememorus*	Int
Celestite (SrSO$_4$)	Actinopoda	Acantharia	Int
		Radiolaria	Int
Barite (BaSO$_4$)	Gamophyta	*Spirogyra*	*Int*
	Charophyta	*Chara*	*Int*
	Rhizopoda	*Xenophyophoria*	*Ext*
"Silica" (opal) SiO$_2 \cdot n$H$_2$O	Actinopoda	Heliozoa, Radiolaria	Int
	Rhizopoda	Widespread in Testacea	Int
	Foraminifera	*Silicoloculina*	Ext
	Zoomastigina	Widespread in loricate choanoflagellates	Int
	Bacillariophyta	Majority	Ext
	Xanthophyta	Small number of coccoid genera	Ext
	?Eustigmatophyta	?Chlorobotrys (cyst)	?
	Pyrrhophyta	*Ebria, Hermesinum*	Int
	Dinoflagellata	Actiniscaceae	Int
	Chrysophyta	Synuraceae, Aurosphaeraceae, Silicoflagellates	Int
?Magnetite (Fe$_3$O$_4$)	?Chlorophyta	*?Chlamydomonas*	Int
"Ferric oxides"	Foraminifera	Many arenaceous genera	Ext
	Euglenophyta	Siderophylic genera	Ext
	Chrysophyta	*Chrysococcus, Synura*	Ext
"Manganese oxides"	Chlorophyta	*Chlamydomonas*	Ext

^{a} From Lowenstam (1986).
^{b} Ext, extracellular; Int, intracellular.

fera but they consist of an inner organic layer and an outer mineralized envelope. Of these, the "porcellaneous" forms are the most interesting in terms of biomineralization since they seem to consist of a three-dimensional disordered array of crystals in which only a surface veneer shows any preferred orientation (Towe and Cifelli, 1967). This unique form of biomineralization apparently involves the formation of calcite needles in Golgi vesicles and their transport to the site of shell deposition, where they form an irregular assembly. The formation of a new chamber in the shell involves extended pseudopodia which transport material to the site of deposition (Hottinger, 1986).

Many foraminifera contain symbiotic algae such as *Chlamydomonas hedleyi*, which occurs in *Archais angulatus*. In these organisms, both photosynthesis and calcification of the test are enhanced by light. Since the photosynthetic inhibitor DCMU (dichlorophenyl dimethylurea) affects both these processes, Duguay and Taylor (1978) suggested that calcification was driven by carbon dioxide fixation. The same phenomenon was observed by Erez (1983) at high concentrations of DCMU, but by adjusting the concentration he was able to inhibit photosynthesis but facilitate calcification, a fact that led him to suggest that the two processes were not coupled but rather interacted through a common carbon pool. A similar conclusion might also be drawn from the fact that, when new chambers are formed in the shell, symbionts may migrate to this site either before or after mineralization occurs (Hottinger, 1988).

B. Silicification

Besides forming calcareous skeletons, many protozoa secrete similar structures made of silica (SiO_2) (Fig. 6.1). Among the radiolarian protozoa, siliceous spicules are deposited within a compartment formed of thin cytoplasmic extensions, the so-called cytokalyemma. It has been suggested that this envelope of cytoplasm forms either by the growing together of rhizopodial strands to enclose a free space into which the skeleton is secreted (the synthesis model) or by the formation of a cyto-

---

Figure 6.1 (a) Whole valve of *Stephanopyxis*. Single ring of fultoportulae occurs on hemispherical valve (Crawford, 1981). (b) External view of *Actinoptychus* valve with six concave and six convex sections (Crawford, 1981). (c). Valve of *Surirella ovalis*. The valve face is basically flat and the marginal canal raphes are not prominent (Pickett-Heaps *et al.*, 1988). (d) View from the outside of a species of *Coscinodiscus* (Crawford, 1981). (e) Valve of *Triceratium trinitas* var. stricta (Simonsen, 1987). (f) Valve of *Cymbella praerupta* (Simonsen, 1987).

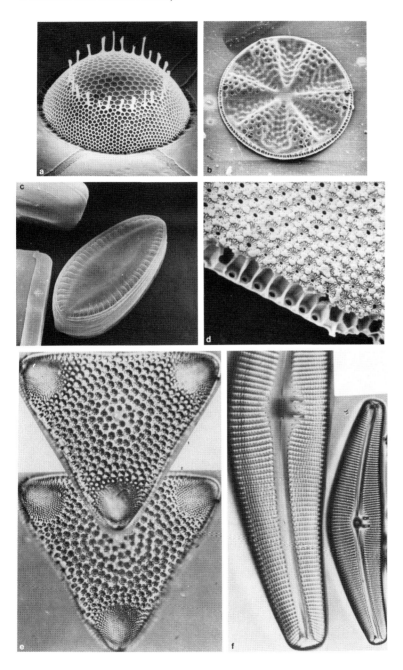

plasmic vacuole from which most of the surrounding cytoplasm withdraws (the segregation model). The net effects would appear very similar (Anderson, 1981). During silicification of the skeleton, numerous small vesicles stream up to the cytoplasmic compartment which, by analogy with other systems, is referred to as a silicalemma. The content of these vesicles is not known but it is assumed that they contain matrix and/or silica components.

The formation of silica ribs or costal strips has been studied in considerable detail in the choanoflagellates (Leadbeater, 1981, 1986). The strips are produced intracellularly in membrane-bound vesicles (silicalemma) from which they are released sideways onto the cell surface. According to Mann *et al.* (1983), such rods show no evidence of any lattice, that is, the silica is truly amorphous. However, it is not a uniform material and it dissolves preferentially from the center to produce hollow tubes. They attribute this to a high degree of hydration which enables the silica to flow out of these tubes. Thus, when two rods are brought together, the inner part may flow out and assist in gluing together the two components of the skeleton.

On a weight basis, silica is the most common material in the Earth's crust. It forms a major component of many rocks and their slow weathering has resulted in a range of concentrations of silicic acid [$SiO_2 \cdot 2H_2O$ or $Si(OH)_4$] in fresh and sea water (up to 180 μM) and in soil water (up to 1.2 μM). Silicic acid dissociates rather poorly and it has an appreciable solubility in the lipids which form the membranes around living cells. After reviewing the biological evidence, Raven (1983) considered that silicic acid had a permeability coefficient of roughly 10^{-10} m^{-1} sec^{-1}. Using this value, he calculated that a marine organism of 10 μm in diameter would be in equilibrium with 100 μM silicic acid in sea water even if it divided several times per day. Thus, because of the intrinsic permeability to silicic acid, organisms are provided with a continual supply of this material. Faced with this fact, cells could presumably extrude it, tolerate it, or utilize it. The energy cost of extruding silicic acid would be very small, perhaps one-thousandth of the cells total energy budget (Raven, 1983) but it would appear to achieve very little. The alternative approach, to turn an available material to a good biological use, appears to have been evolved several times by organisms that form siliceous skeletons. There are, however, a number of interesting biological factors that have probably influenced this process. The first is that, in order to exceed the solubility product, the silicic acid would have to be concentrated to a level of about 2 mM, i.e., 10 or more times the level of most aquatic systems. A second problem for mobile aquatic cells is that

SiO_2, even when hydrated, is quite dense and when used as a skeleton to any significant extent it is liable to cause the organism to sink in the water column. Finally, silica is a relatively soluble mineral that will redissolve in the water unless protected by some organic material. On the other hand, there are advantages to silica as a skeletal material. Thus, it is an amorphous material that is easily shaped by the cell (Simkiss, 1986) and it tends not to crystallize to quartz because of a large kinetic barrier (Williams, 1981). It is also energetically a rather cheap material to use for structural and defensive purposes since it has been calculated that it would cost the cell only 5 or 10% of the energy that would be required to form an organic skeleton of the same mass (Raven, 1983).

The organisms that have perfected silicification are, of course, the diatoms (Chrysophyta, Table 6.1). They live in a rigid siliceous exoskeleton composed of two parts that overlap, like a box (hypotheca) and its lid (epitheca). When cell division occurs, each daughter cell inherits one part of the skeleton which it uses as an epitheca and within which it forms a new hypotheca (Schmid, 1987). As a result, those descendants that continually inherit the "box" component and use it as a "lid" get progressively smaller and smaller until sexual reproduction allows the process to return to normal. This is a particularly dramatic demonstration of the constraints of an exoskeleton on the growth of an organism. It also demonstrates how diatoms are dependent on a supply of silica in order to grow. In fact, silicic acid is accumulated in these organisms by carrier-mediated systems and Bhattacharyya and Volcani (1980) have used both intact cells and membrane vesicles to demonstrate the existence of a sodium-dependent system that appears to be based on a Na^+/silicic acid symport mechanism.

Intracellular silicification occurs in diatoms, chrysophytes, choanoflagellates, radiolarians, and testaceous amoebae (Table 6.1). The process occurs within a membrane-deliminated region, the silicalemma. In the diatoms, the mineralized wall is composed of a siliceous portion, the frustule, with a firmly bound organic casing. It is formed in a "silica deposition vesicle" whose origin has variously been ascribed to the Golgi system or the fusion of a number of smaller vesicles. The walls of the vesicle are thought to act as nucleation sites and the chemical form of the intracellular silicon pool changes during silicification so that a trichloracetic acid-extractable fraction increases as the process continues (Robinson and Sullivan, 1987). According to Schmid and Schulz (1979), small electron-dense vesicles (30–40 nm diameter) coalesce to form the silicalemma while at the same time they release their siliceous contents.

When the deposit is fully formed, it fuses with the plasmalemma and silicalemma to form a new cell wall, while another plasmalemma forms beneath it (Li and Volcani, 1984).

This interpretation, implying a membrane-bound system of vesicles as the basis of silicification has, however, been challenged on the basis of electron spectroscopic imaging of the diatom *Thalassiosira pseudonana*. According to Rogerson *et al.* (1987), silica is not concentrated in vesicles but occurs with the ribosomes, suggesting that it is involved in a more direct synthetic role in the cell's metabolism.

The other major group of siliceous organisms is the golden algae or Chrysophyceae. In contrast to the diatoms, the silica deposition vesicle is fully formed prior to silica deposition and it has therefore been suggested that this organelle acts as a mold which shapes the subsequent mineral deposit (McGrory and Leadbeater, 1981). In *Synura* there is no organic layer equivalent to the coating seen in diatom frustules and the silicalemma does not become incorporated into the final scale (Leadbeater, 1984).

II. Calcification and Coccolithophorids

The coccolithophorids are algae of the class Prymnesiophyceae that form species-specific plates of $CaCO_3$. As members of the oceanic phytoplankton, coccolithophorids have an important place in the economy of the sea, where they act as substantial contributors to the food chains that support zooplankton. Their distribution includes tropical, subtropical, subarctic, and subantarctic waters (McIntyre and Bé, 1967). The highest concentrations are found in surface waters down to 100 m, and lower concentrations occur down to depths below 4000 m (Okada and Honjo, 1973). Some coccolithophorids are present in freshwater.

Since the form of the $CaCO_3$ structures produced by each species is distinctive, it has been possible to identify species in geological strata. Coccoliths were present in the early Jurassic and they continued in abundance until the end of the Cretaceous when most, but not all, the calcareous nanoplankton disappeared (Haq, 1978). Because of the numbers of coccolithophorids in the past geological periods and their prodigious ability to convert calcium and bicarbonate ions to $CaCO_3$, this group of algae has been responsible for forming extensive beds of $CaCO_3$ in the Earth's crust.

The Coccolithophoridae is one of the many groups of plants and invertebrates that deposit mineral intracellularly within vacuoles. However, mineralization by the coccolithophorids differs from that of many organisms in that a complex mineral structure is produced and precisely assembled within a single Golgi vesicle. The secreted structures are usually known as coccoliths and once formed they are extruded through the cell membrane and encase the cell (Figs. 6.2 and 6.3A,B). As a consequence, the sequence of steps by which coccoliths are constructed has long attracted microscopists, biochemists, and physiologists. (Klaveness and Paasche, 1979).

A. Structure of Cells, Scales, and Coccoliths

External scales on the surface of unicellular algae were one of the early spectacular discoveries revealed by electron microscopy. The pioneering

Figure 6.2 Scanning electron micrograph of *Cyclococcolithus leptoporus* showing the numerous coccoliths surrounding the cell. The overlapping individual calcite crystals making up the coccoliths can be seen [Borowitzka (1982), after Dr. G. Hallegraeff].

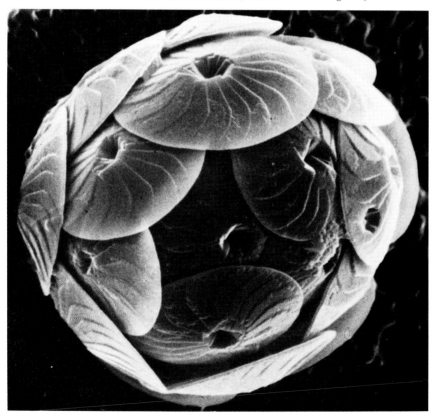

investigations of Parke and Manton (1962) showed a variety of simple plates, elaborate scales, and ornate spines. The basic pattern appeared to be a two-layered plate with a pattern of radial fibrils on the face directed toward the cell and a pattern of spirally oriented fibrils on the

Figure 6.3 (A) Cross-section through calcifying cell of *Emiliania huxleyi*. Drawing based on micrographs of ultrathin sections of specimens fixed in glutaraldehyde and OsO$_4$. Within the coccolith vesicle there is an immature coccolith. Only two of the extracellular coccoliths constituting the coccosphere are shown. Coccoliths are composed of about 30 units of radially arranged crystalline units, each of which can be subdivided into a connecting wall (a) between a lower element (b) and an upper element (c). The central part of the intracellular coccolith consists of a plate-like structure (arrow). Abbreviations: Ch, chromatin; Chl, chloroplast; Cov, cover; CV, coccolith vesicle; EC, extracellular coccolith; G, Golgi complex; M, mitochondrion; N, nucleus; NE, nuclear envelope; RB, reticular body; V, cell vacuole; X, crystalline matter (van der Wal *et al.*, 1983b). (B) A coccolith of the calcareous marine alga *Emiliania huxleyi*. Carbon replica, platinum shadowed. (Courtesy of S. Collings.) (*Figure continues.*)

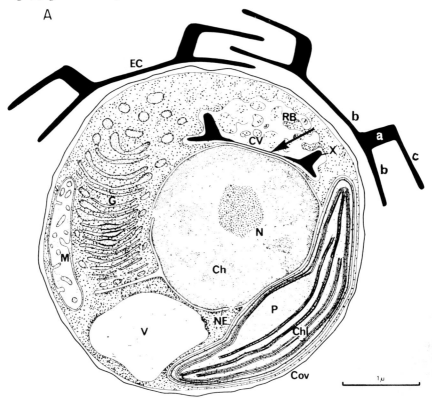

Figure 6.3 (B) (*continued*)

distal face. The scales are basically composed of polysaccharide material with a little protein and are produced within the cisternae of the Golgi body in a complex series of transformations (Green, 1986). The importance of such scales in discussions of coccolith formation is threefold. First, most coccoliths are formed on such an organic base plate or scale.

Second, many coccolithophorids may produce coccoliths and scales within the same cell either simultaneously, as in *Hymenomonas carterae* (= *Pleurocrysis carterae*) (van der Wal *et al.*, 1983a), or at various stages of the life cycle, as in *Emiliania huxleyi* (Klaveness and Paasche, 1979). Third, the formation of scales and superimposed calcified plates occurs within the same organelles, namely the Golgi vesicles.

The two species used most commonly in current research are *E. huxleyi* and *H. carterae*. The beautifully sculptured coccoliths of *E. huxleyi* consist of two disks joined by a central cylinder. The upper disk is made up of many spokes, each with a hammer head (Fig. 6.3B). The coccoliths of *H. carterae* are oval rings of anvil-shaped interlocking mineral units of two alternating shapes (Outka and Williams, 1971).

The cells which produce coccoliths usually have one or two prominent chloroplasts, several mitochondria, and a membrane system consisting of rough endoplasmic reticulum closely associated with a very-well-developed Golgi complex in which the coccoliths are formed (Fig. 6.4). Bundles of microtubules may be present and have been considered to have various functions, including the extrusion of coccoliths through the plasmalemma (Brown and Romanovicz, 1976).

B. Coccolith Development

The sequence of steps in the formation of coccoliths has been described by several authors (e.g., Manton and Leedale, 1969; Outka and Williams, 1971; van der Wal *et al.*, 1983a,b). Our comments on development will, therefore, give primary attention to those aspects which pertain to the deposition of calcium carbonate. The structures immediately associated with mineralization of the coccolith are (1) a Golgi vesicle bounded by a membrane through which ions must pass inward from the cytosol; (2) a scale formed within the cisterna on which the coccolith will be deposited and attached, although an exception is *E. huxleyi* in which the scale and coccoliths are formed independently; and (3) an organic matrix within which the coccolith develops.

The pattern of development of coccoliths will vary, of course, with the coccolith morphology of the particular species. In picturing the development, it should be remembered that the coccoliths of many species are complex structures of many mineral units. For example, a coccolith of *H. carterae* has some 30 units in its oval ring (van der Wal *et al.*, 1983a), and *Coccolithus pelagicus* built on a similar plan is made up of 42 to 52 units (Manton and Leedale, 1969). To form such coccoliths requires a precisely spaced distribution of crystal nucleation sites within the matrix of the

Figure 6.4 Coccolith formation within Golgi vesicle of *Hymenomonas*. (A) Early stage: scale on which coccolith will be formed. The coccolithosomes are the dark circles in the "ears" of vesicle. The vacuole has the general shape of the final form of a coccolith in vertical section. (B) Later stage; region of beginning mineralization is indicated by bars at left and right. The coccolithosomes are the dark circles (Outka and Williams, 1971).

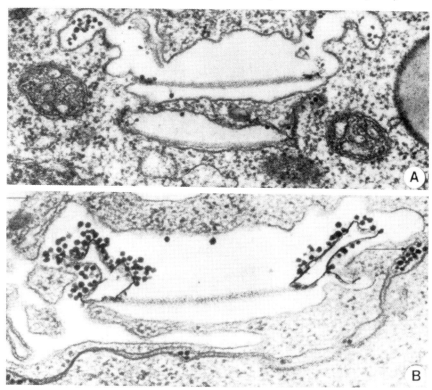

Golgi cisterna. From these sites the individual mineral units will follow a species-specific pattern to give an integrated coccolith. In *E. huxleyi* the formation of each of the mineral units begins at its base (Fig. 6.4) (Wilbur and Watabe, 1963; Outka and Williams, 1971). Vertical progression of mineralization of the units forms the central cylinder and outward development of each mineral unit results in the elements of the upper disk. Mineralization of the base of each unit, both outward and centrally produces the lower disk or base plate. The 30 or so individual units, by joining or becoming contiguous, form the completed coccolith.

There are several factors involved in the whole process. Manton and Leedale (1969) observed a noncalcified organic replica of a coccolith of *C.*

pelagicus which closely resembled the mineralized form. They commented that this provided "exceptionally clear evidence for the importance of the matrix in determining coccolith morphology." One might, perhaps, have also added the importance of the vesicle in further determining this shape. Watabe (1967) was the first to study a single unit of a coccolith and show that the lower element behaved as a single crystal. More recently, Parker *et al.* (1983) have used ultrahigh-resolution electron microscopy to probe the structure of the upper portion and suggested that it consists of a composite of microdomains no more than 30–50 nm in size. The relevance of all these features will, however, only be apparent when we appreciate the different mechanisms of coccolith formation.

Central to the secretory activities of most cells is the Golgi apparatus. This consists of an oriented stack of disk-shaped cisternae. On one surface (the *cis* face) they receive coated vesicles from the endoplasmic reticulum, whose secretory products pass into the lumen of the Golgi system. After suitable packaging, which often appears to involve changes to the carbohydrate components, these products are exported from the concave surface (or *trans* face) as vesicles. These subsequently pass to various organelles which, in the case of the coccolithophorids, include the sites of coccolith secretion.

The marine alga *H. carterae* produces both scales and calcified coccoliths simultaneously within the membranes of the Golgi system (Outka and Williams, 1971). The general ultrastructure of the cell contains a prominent Golgi body with something like 16 easily identified cisternae close to the nucleus. The system, as it is envisaged by van der Wal *et al.* (1983a), is shown diagramatically in Fig. 6.5. They believe that, unlike the system in most cells, the Golgi vesicles migrate from the main body of the apparatus and move toward the cell boundary. During this process they receive material from vesicles that is used to form the organic scales. The formation of the calcified coccoliths is more complex. Special vesicles containing electron-dense bodies (the so-called coccolithosomes, Fig. 6.4) are produced from the Golgi system and pass through the cytoplasm to fuse with coccolith vesicles. It is thought that the coccolithosomes carry both calcium and matrix material to these sites of calcification. van der Wal *et al.* (1985) measured the calcium content of a coccolithosome by X-ray microanalysis and calculated that 70,000 coccolithosomes are needed to form one coccolith. Since several coccoliths may be formed per hour, several hundred thousand coccolithosomes may be produced by one Golgi body in 1 hr. The developing scales and coccoliths are easily identified in the more peripheral parts of the cyto-

Figure 6.5 Cross-sectional model of scale and coccolith biosynthesis in *Hymenomonas carterae*. The specific stages in the assembly process are represented. The pile of tightly stacked cisternae can be divided into a proximal and a distal zone with the cisternae marked IC in an intermediate position. Also indicated are organic covering of the crystallites (coat), chloroplast (Chl), columnar material (CM), coccolithosomes (cs), extracellular coccolith (EC), endoplasmic reticulum (ER), nucleus (N), plasma membrane (PM), and a scale (sc) (van der Wal *et al.*, 1983a).

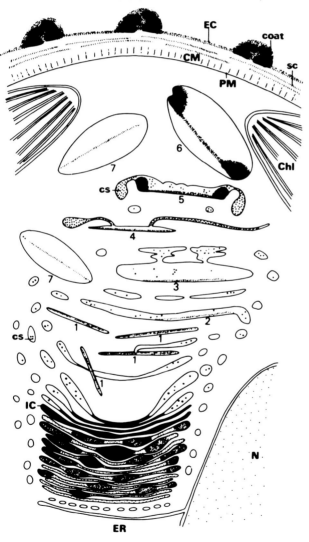

plasm, and eventually they are exocytosed onto the surface of the cell. The coccoliths are stuck together to form an outer layer beneath which are the organic scales (but not in E. *huxleyi*) and an amorphous "columnar material" on the outer cell membrane (Figs. 6.2 and 6.3A). It is thought that there will be a continual recycling of the membrane components of the vesicles and Golgi system during this secretory process.

In the coccolithophorid E. *huxleyi,* the coccolith is formed in a particular vesicle closely associated with the nucleus (Fig. 6.3). Attached to the coccolith vesicle is a reticular body consisting of an anastomosing series of tubes (Wilbur and Watabe, 1963). It is thought that the reticular body is also responsible for passing matrix material and possibly calcium, which it receives from Golgi vesicles, to the forming coccolith. Once the coccolith is formed, the reticular body detaches and degenerates while the coccolith vesicle moves to the surface of the cell and exocytoses the mineralized body (Westbroek *et al.*, 1986). The association of the coccolith vesicle and the Golgi system is, perhaps, not as clear in E. *huxleyi* as in other algae but the vesicle always forms close to the Golgi, which also stains for carbohydrate and calcium ions.

C. Organic Matrix Associated with Coccoliths

The organic matrix material within which coccoliths are formed is thought to have a possible role in governing their detailed morphological development. For this reason, matrix properties have been given particular attention, especially by the group at the University of Leiden (de Vrind-de Jong *et al.*, 1986; Borman *et al.*, 1982). The term matrix as used in coccolithophorids may refer to the organic material which fills the Golgi vesicles or, more specifically, to the material immediately surrounding the developing coccoliths and which remains attached after the completely formed coccoliths are extruded to the cell surface. Matrix has also been demonstrated within crystals forming the individual mineral units of coccoliths (N. Watabe, personal communication). The matrix of E. *huxleyi,* which partly surrounds coccoliths after their extrusion, differs from the matrix of most calcified structures in that it does not seem to contain a water-insoluble fraction. The water-soluble matrix which is present contains 13 different monosaccharides arranged as a highly branched mannose backbone containing ester sulfate groups and with uronic acids and methylated sugars located in the periphery (de Vrind-de Jong *et al.*, 1986).

The matrix from coccoliths, in common with similar polysaccharides from other mineralized deposits, has the ability to delay the precipitation of $CaCO_3$ *in vitro* (Borman *et al.*, 1983). It probably binds, via the uronic acid groups, to clusters of $CaCO_3$ as well as to growth sites on the crystals and prevents the growth of these nuclei. This property was used by Westbroek *et al.* (1984) as an ingenious explanation for the function of the Golgi vesicle. In this hypothesis, the matrix polysaccharide extends through the "reticular body–Golgi vesicle" system and inhibits mineral formation. When the vesicle expands and the polysaccharide becomes attached to the base plate it ceases to inhibit and may in fact facilitate crystal growth. The first crystallites are therefore formed on the rim of the base plate and they continue to grow as the cytoskeleton pulls on the walls of the vesicle and dictates its shape. Eventually, a layer of polysaccharide covers the coccolith, causing its growth to stop. Thus, at least three functions are attributed to the polysaccharide in this hypothesis. Initially, in the flattened vesicle, it is inhibitory. When the vesicle expands, the polysaccharide attaches to protein and induces crystal growth. Finally, when the polysaccharide is present as a free molecule it again covers the coccolith and inhibits further growth.

This explanation of coccolith formation accounts for the initiation and cessation of crystal growth but it is still necessary to account for the formation of a large number of crystal units of precise and complex shape in a coordinated way. Watabe (personal communication) suggests the involvement of local inhibition of crystal growth, presumably by the matrix, in producing the units of characteristic form. Such a system would obviously reflect the genetic control of the organism over the specific shape of its coccoliths.

D. Ion Movements

The calcium ions necessary for the formation of coccoliths can be expected to move from the medium into the cell by diffusion. The movement will be favored by a potential difference since the inside of the cell is negative (Sikes and Wilbur, 1982), and by a very low calcium ion activity in the cytosol. By using cytochemical methods and X-ray microanalysis, van der Wal *et al.* (1985) identified a high concentration of calcium in the Golgi system, suggesting that this organelle concentrates the ion from the cytoplasm and then presumably transfers it to the coccolith vesicle via coccolithosomes or other Golgi vesicles. The input of energy necessary to accomplish this initial accumulation may well come from an ATPase, but whether a Ca^{2+}-ATPase, a H^+-ATPase, or some other type of transport is involved is unknown. A H^+-ATPase, demonstrated for other plant vacuoles (page 110 and Fig. 8.1), might also move H^+ outward from the coccolith vesicle increasing its pH and so favoring calcium carbonate deposition. The possible involvement of these ATPase enzymes is indicated by the reversible inhibition of coccolith formation by oligomycin, a type F ATPase inhibitor (Dorigan and Wilbur, 1973). The movement of bicarbonate ions into vesicles as a possible source of carbonate has not been studied. Our understanding of the mechanisms of ion transport would therefore be greatly facilitated if vesicles could be isolated for *in vitro* experimentation by using the methods that have been successfully carried out with vacuoles of higher plants.

In *Crystallolithus*, a genus of the Prymnesiophyceae, there appear to be two calcification systems, one internal and one external. The first involves the formation of coccoliths within Golgi vesicles as previously described. In the other, coccoliths called holococcoliths develop external to the plasmalemma within a microspace bounded by a secreted outer investment called the "skin" (Green, 1986). It appears that the holococcoliths are formed by the active transport of calcium outward through the plasmalemma into this microspace. The "skin" presumably excludes the surrounding medium and so permits the transported Ca^{2+} to exceed the solubility product of the fluid within the microspace and so the holococcoliths are formed.

The morphology and mineralogy of coccoliths can be changed experimentally. By growing cultures of *E. huxleyi* at constant temperatures over a range from 7° to 27°C, it was found that the precise form of coccoliths is temperature dependent (Watabe and Wilbur, 1966). The effect is sufficiently distinct to distinguish between populations grown at temperatures differing by 3°–6°C. In addition to influencing the shape of coccoliths, temperature also affected their rate of formation. The percentage of cells forming coccoliths was 2 to 3-fold greater at 18° and 24° than at 7° and 1°C.

The mineral deposited in coccoliths is calcite. In *H. carterae*, decreasing the concentration of magnesium ions in the culture medium changed this to 60% calcite and 40% aragonite (Blackwelder *et al.*, 1976). This effect is not easily explained. Certain organic compounds and ions are known to alter the calcite–aragonite ratio of calcium carbonate formed in solution (Kitano *et al.*, 1969), but what changes occur within the cell from a low concentration of magnesium in the medium are unknown. Since magnesium normally induces aragonite formation and since it is in high concentration in the culture medium, it is surprising that a reduction caused aragonite formation.

E. Photosynthesis and Mineralization

It has long been known that light and photosynthesis facilitate the deposition of calcium carbonate, but the mechanism by which the effect is produced is still to be clarified. Part of the problem is that the experimental results have not been uniform. Some investigators have demonstrated that light is required (Crenshaw, 1964; Dorigan and Wilbur, 1973) whereas others (Blankley, 1971; Ariovich and Pienaar, 1979; van der Wal *et al.*, 1984) have found that coccoliths are formed in the absence of light. One experimental approach has been to expose *H. carterae* to a 16 : 8 light–dark cycle and follow the rate of calcium uptake by sampling at frequent intervals over a period of 51 hr (van der Wal *et al.*, 1984). Under these conditions, the rate of calcium deposition was unchanged during the periods of darkness. However, in constant darkness, the rate was markedly slowed (Fig. 6.6B). If the limiting factor in an extended period of darkness is the energy supply from photosynthesis, it would appear that within a 16-hr light period, enough ATP is formed to provide energy for coccolith synthesis for an 8-hr period without light (Fig. 6.6A).

Figure 6.6 (A) Calcium accumulation in extracellular coccoliths (open circles) and in the intracellular pool (closed circles) of *Pleurochrysis carterae.* Shaded areas are periods of darkness; unshaded areas are periods of light. (B) Accumulation of calcium in the extracellular pool, i.e., extracellular coccoliths (open circles), and in the intracellular pool (closed circles) of cells of *P. carterae* cultured in the dark in a medium to which calcium-45 had been added at 0 hr. Figure represents an experiment with nondecalcified cells (van der Wal *et al.*, 1984).

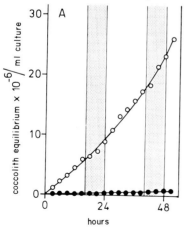
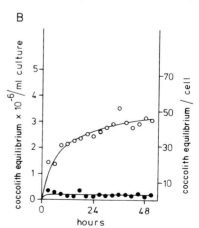

The first question to be answered in the linking of photosynthesis and $CaCO_3$ deposition is the carbon species entering the cell for each process. This has been examined in *E. huxleyi* and *H. carterae* by following the rates of incorporation of $^{14}CO_2$ and $H^{14}CO_3^-$ into the photosynthate and calcium carbonate of the coccoliths (Sikes *et al.*, 1980). The amounts of isotope incorporated were determined at intervals over a 2-min period before they equilibrated in the medium (CO_2, $t_{1/2} = 20$ sec; HCO_3^-, $t_{1/2} = 15$ sec). Fig. 6.7 shows the interpretation of the data. A similar scheme is presented by Borowitzka (1982). HCO_3^- is the form of carbon provided in calcification (A); CO_2 is the substrate of photosynthesis (B,C,E): CO_2 and HCO_3^- is the form of carbon provided in calcification (A); CO_2 and HCO_3^- entering the cells from the medium are used in photosynthesis (E,B); CO_2 also removed proton (A,B) and could accordingly favor continued calcification. However, photosynthesis will not be required pro-

Figure 6.7 Possible reactions relating to coccolith formation and photosynthesis. The function of carbonic anhydrase (CA, reaction B) is assumed (after Sikes *et al.*, 1980).

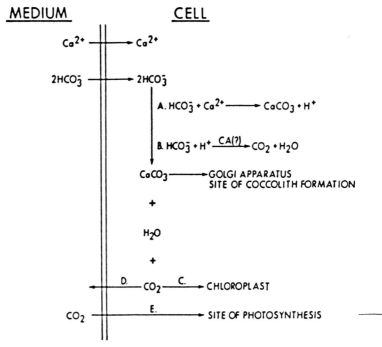

vided that the CO_2 diffuses sufficiently rapidly from the site of mineral deposition. Apparently, this takes place in corals at ocean depths where light is absent. A factor required for $CaCO_3$ formation in addition to proton removal is the production of ATP, which provides energy for ion transport and the synthesis of organic matrix. The absence of photosynthesis in darkness may also decrease the rate of $CaCO_3$ due to reduction in ATP production.

III. Summary

1. Because of their availability and ease of handling, the mineralization systems of the protoctistan phyla have attracted considerable interest.
2. Among the protozoa some mineral deposits may be simply stores of metabolites such as phosphate. Other minerals such as the crystals of strontium sulfate in radiolarians or the barium sulfate deposits of desmids probably reflect the activity of specific ion pumps.
3. The foraminifera produce a great variety of calcareous structures. Those species which contain symbiotic algae show an enhancement of calcification during algal photosynthesis.
4. Many protozoa also secrete siliceous skeletons. Details of the process vary but they invariably involve a membrane-enclosed vesicle in which the silica is deposited.
5. A similar silica deposition vesicle occurs in diatoms and there is good evidence to show that active transport processes are involved with the uptake of silicic acid.
6. A number of single-celled algae produce calcareous plates. The most extensive studies of these plants are on the coccolithophorids which form species-specific plates or coccoliths. The ultrastructural and biochemical aspects of this process have been studied in detail. Upon completion, these coccoliths are extruded from the Golgi vesicle and become attached to the outer cell surface.
7. The formation and assembly of a coccolith, which may consist of many units of complex form, require accurate spacing of an equal number of nucleation sites from which the units are initiated. It appears likely that the control of the ultrastructure is under genetic control and the expression of this invovles the organic matrix.
8. The matrix material includes carbohydrate molecules containing uronic acid with carboxyl groups which bind calcium.
9. The calcium and bicarbonate ions which form the calcite of the coc-

coliths enter the cytosol from the medium and are then transported across the membrane of the Golgi cisternae. Details of the mechanisms of transport are not known, but it is certain that an active transport is required for moving calcium from a low cytosol concentration to one which exceeds the solubility product of calcium carbonate within cisternae. The mechanisms of transport could be studied to advantage *in vitro*, provided it were possible to isolate cisternae in the process of mineral deposition.

10. Recent studies have demonstrated that calcium carbonate can be deposited in the absence of light for a period of several hours, but that the rate of deposition is markedly decreased with extended periods of darkness. Light and photosynthesis probably facilitate coccolith production by providing ATP as an energy source for matrix synthesis and the transport of matrix and ions across the cisternal membrane.

11. The intricacy of many protoctistan skeletons is probably the reflection of three important features. First, siliceous skeletons are formed of amorphous silica. There is no long-range order to these deposits so that they can be shaped by biological "molds." Second, some of the calcareous skeletons appear to be mixtures of well-oriented crystals with areas composed of microdomains. Parts of the skeleton where crystallites are very small can be oriented to form intricate patterns independent of the constraints of the crystal lattice. Finally, the crystal can itself be shaped. Certain crystal faces can be inhibited from growing by specific poisons that attach to those particular lattice spacings; others can be shaped by the physical interference of matrix proteins and vesicle membranes or by the supply of ions from only one direction. These are all exciting aspects of molecular biology and their effects are most clearly seen in the beauty of the protoctistan skeletons.

References

Anderson, O. R. (1981). Radiolarian fine structure and silica deposition. *In* "Silicon and Siliceous Structures in Biological Systems" (T. L. Simpson and B. E. Volcani, eds.), pp. 347–379. Springer-Verlag, New York.

Anderson, O. R., and Bé, A. W. H. (1978). Recent advances in foraminiferal fine structure research. *Foraminifera* **3**, 121–202.

Ariovich, D., and Pienaar, R. N. (1979). The role of light in the incorporation and utilization of Ca^{++} ions by *Hymenomonas carterae* (Braarud et Fagerl.) Braarud (Prymnesiophyceae). *Br. Phycol. J.* **14**, 17–24.

Bhattacharyya, P., and Volcani, B. E. (1980). Sodium-dependent silicate transport in the apochlorotic marine diatom *Nitzchia alba. Proc. Natl. Acad. Sci. U.S.A.* **77,** 6386–6390.

Blackwelder, P. L., Weiss, R. E., and Wilbur, K. M. (1976). Effects of calcium, strontium and magnesium on the coccolithophorid *Cricosphaera (Hymenomonas) carterae.* I. Calcification. *Mar. Biol.* **34,** 11–16.

Blankley, W. F. (1971). "Auxotrophic and Heterotrophic Growth and Calcification in Coccolithophorids," Ph.D. dissertation. Univ. of California, San Diego, California.

Borman, A. H., de Jong, E. W., Huizinga, M., Kok, D. J., Westbroek, P., and Bosch, L. (1982). The role of CaCO₃ crystallization of an acid Ca²⁺-binding polysaccharide associated with coccoliths of *Emiliania huxleyi. Eur. J. Biochem.* **129,** 179–183.

Borman, A. H., de Jong, E. W., Huizinga, M., and Westbroek, P. (1983). Inhibition of CaCO₃ precipitation of a polysaccharide associated with coccoliths of *Emiliania huxleyi. In* "Biomineralization and Biological Metal Accumulation" (P. Westbroek and E. W. de Jong, eds., pp. 303–305. Reidel, Dordrecht, The Netherlands.

Borowitzka, M. A. (1982). Mechanisms in algal calcification. *Prog. Phycol. Res.* **1,** 137–177.

Brown, R. M., and Romanovicz, D. K. (1976). Biogenesis and structure of Golgi-derived cellulosis scales in scale assembly and exocytosis. *Appl. Polym. Symp.* **28,** 537–585.

Crawford, R. M. (1981). The siliceous components of the diatom cell wall and their morphological variation. *In* "Silicon and Siliceous Structures in Biological Systems" (T. L. Simpson and B. E. Volcani, ed.), pp. 129–156. Springer-Verlag, Berlin.

Crenshaw, M. A. (1964). "Coccolith Formation by Two Murine Coccolithophorids, *Coccolithus huxleyi* and *Hymenomonas sp,*" Ph.D. thesis. Duke University Durham, North Carolina.

de Vrind-de Jong, E. W., Borman, A. H., Thierry, R., Westbroek, P., Gruter, M., and Kamerling, J. P. (1986). Calcification in the coccolithophorids *Emiliania huxleyi* and *Pleurochrysis carterae.* II. Biochemical aspects. *Syst. Assoc. Spec. Vol.* **30,** 206–217.

Dorigan, J. L., and Wilbur, K. M. (1973). Calcification and its inhibition in coccolithophorids. *J. Phycol.* **9,** 450–456.

Duguay, L., and Taylor, D. L. (1978). Primary production and calcification by the soritid foraminifer *Archais augulatus. J. Protozool.* **25,** 356–361.

Erez, J. (1983). Calcification rates, photosynthesis and light in planktonic foraminifera. *In* "Biomineralization and Biological Metal Accumulation" (P. Westbroek and E. W. de Jong, eds.), pp. 307–312. Riedel, Dordrecht, The Netherlands.

Faure-Fremiet, E. (1957). Concrétions minérales intracytoplasmiques chez les cilies. *J. Protozool.* **4,** 96–109.

Faure-Fremiet, E., Andre, J., and Ganier, M. C. (1968). Calcification tegumentaire chez les cilies du genre Coleps Nitzsch. *J. Microsc.* **7,** 693–704.

Green, J. C. (1986). Biomineralization in the algal class Prymnesiophyceae. *Syst. Assoc. Spec. Vol.* **30,** 173–188.

Hanson, E. D. (1967). Protozoan development. *Chem. Zool.* **1,** 395–539.

Haq, B. U. (1978). Calcareous nanoplankton. *In* "Introduction to Marine Micropaleontology" (B. V. Haq and A. Boersma, eds.), pp. 79–108. Elsevier, New York.

Hollande, A., and Martoja, R. (1974). Identification du cristalloide des isospores de radiolaires a un cristal de celestite (SrSO₄). Determination de la constitution du cristalloides per voie cytochemique et a l'aide de la microsonde electronique et du microanalyseur a emission ionique secondaire. *Protistologie* **10,** 603–609.

Holligan, P. M., Viollier, M., Harbour, D. S., Camus, P., and Champagne-Phillipe, M. (1983). Satellite and ship studies of coccolithophore production along a continental shelf edge. *Nature (London)* **304,** 339–342.

Hottinger, L. (1986). Construction, structure and function of foraminiferal shells. *Syst. Assoc. Spec. Vol.* **30**, 219–249.

Jones, A. R. (1967). Calcium and phosphorus accumulation in *Sirostomum ambiguum*. *J. Protozool.* **14**, 220–225.

Jones, A. R., Taylor, M. G., and Simkiss, K. (1984). Regulation of calcium, cobalt and zinc by *Tetrahymena elliotti*. *Comp. Biochem. Physiol. A* **78**, 493–500.

Kitano, Y., Kanamori, N., and Tokuyama, A. (1969). Effects of organic matter on solubilities and crystal formation of carbonates. *Am. Zool.* **9**, 681–688.

Klaveness, D., and Paasche, E. (1979). Physiology of coccolithophorids. *In* "Biochemistry and Physiology of Protozoa" (M. Levandowsky and S. H. Hunter, eds.), Vol. 1, 2nd Ed. pp. 191–213. Academic Press, New York.

Leadbeater, B. S. C. (1981). Ultrastructure and deposition of silica in loricate choanoflagellates. *In* "Silicon and Siliceous Structures in Biological Systems" (T. L. Simpson and E. Volcani, eds.), pp. 295–322. Springer-Verlag, Berlin.

Leadbeater, B. S. C. (1984). Silicification of "cell walls" of certain protistan flagellates. *Philos. Trans. R. Soc. London, Ser. B* **304**, 529–536.

Leadbeater, B. S. C. (1986). Silica deposition and lorica assembly in choanoflagellates. *Syst. Assoc. Spec. Vol.* **30**, 345–360.

Li, C. V., and Volcani, B. E. (1984). Aspects of silicification in wall morphogenesis of diatoms. *Philos. Trans. R. Soc. London, Ser. B.* **304**, 519–528.

Lowenstam, H. A. (1986). Mineralization processes in monerans and protoctists. *Syst. Assoc. Spec. Vol.* **30**, 1–17.

McGrory, C. B., and Leadbeater, B. S. C. (1981). Ultrastructure and deposition of silica in the Chrysophyceae. *In* "Silicon and Siliceous Structures in Biological Systems" (T. L. Simpson and E. Volcani, eds.), pp. 201–230. Springer-Verlag, Berlin.

McIntyre, A., and Bé, A. W. H. (1967). Modern coccolithophorids of the Atlantic Ocean— I. Placoliths and cyrtoliths. *Deep-Sea Res.* **14**, 561–597.

Mann, S., Parker, S. B., Perry, C. C., Ross, M. D., Skarnulis, A. J., and Williams, R. J. P. (1983). Problems in the understanding of biominerals. *In* "Biomineralization and Biological Metal Accumulation" (P. Westbroek and E. W. de Jong, eds.), pp. 171–183. Riedel, Dordrecht, The Netherlands.

Manton, I., and Leedale, G. F. (1969). Obervations on the microanatomy of *Coccolithus pelagicus* and *Cricosphaea carterae*, with special reference to the origin and nature of coccoliths and scales. *J. Mar. Biol. Assoc. U.K.* **49**, 1–16.

Margulis, L., and Stolz, J. F. (1983). Microbial systematics and a Gaian view of the sediments. *In* "Biomineralization and Biological Metal Accumulation" (P. Westbroek and E. W. de Jong, eds.), pp. 27–53. Riedel, Dordrecht, The Netherlands.

Nilsson, J. R., and Coleman, J. R. (1977). Calcium rich refractile granules in *Tetrahymena pyriformis* and their possible role in intracellular ion regulation. *J. Cell Sci.* **24**, 311–325.

Okada, H., and Honjo, S. (1973). The distribution of oceanic coccolithophorids in the Pacific. *Deep-Sea Res.* **20**, 355–374.

Outka, D. E., and Williams, D. C. (1971). Sequential coccolith morphogenesis in *Hymenomonas carterae*. *J. Protozool.* **58**, 285–297.

Parke, M., and Manton, I. (1962). Studies on marine flagellates VI. *Chrysochromulina pringsheimii* sp. nov. *J. Mar Biol. Ass. U.K.* **42**, 391–404.

Parker, S. B., Skarnulis, A. J., Westbroek, P., and Williams, R. J. P. (1983). The ultrastructure of coccoliths from the marine alga *Emiliania huxleyi* (Lohman). *Proc. R. Soc. London, Ser. B* **219**, 111–117.

Pautard, F. G. E. (1970). Calcification in unicellular organisms. In "Biological Calcification" (H. Schraer, ed.), pp. 105–201. Appleton-Century-Crofts, New York.

Perry, C. C., Wilcock, J. R., and Williams, R. J. P. (1988). A physicochemical approach to morphogenesis: the roles of inorganic ions and crystals. Experientia, 44, 638–650

Pickett-Heaps, J. D., Cohn, S., Schmid, A.-M. M., and Tippet, D. H. (1988). Valve morphogenesis in Surirella. J. Phycol. 24, 35–49.

Raven, J. A. (1983). The transport and function of silicon in plants. Biol. Rev. 58, 179–207.

Rieder, N., Ott, H. A., Pfundstein, P., and Schock, R. (1982). X-Ray microanalysis of the mineral contents of some protozoa. J. Protozool. 29, 15–18.

Robinson, D. H., and Sullivan, C. W. (1987). How do diatoms make silicon biominerals? Trends Biochem. Sci. 12, 151–154.

Rogerson, A., Defreitas, A. S. W., and McInnes, A. G. (1987). Cytoplasmic silicon in the centric diatom Thalassiosira pseudonana localized by electron spectroscopic imagining. Can. J. Microbiol. 33, 128–131.

Rosenberg, H. (1966). The isolation and identification of "volutin" granules from Tetrahymena. Exp. Cell Res. 41, 397–410.

Schmid, A. M. (1987). Morphological forces in diatom cell wall formation. In "Cytomechanics" (J. Bereiter-Hahn, O. R. Anderson, and W. Reif, eds., pp. 183–199. Springer-Verlag, Berlin.

Schmid, A. M., and Schulz, D. (1979). Wall morphogenesis in diatoms: Deposition of silica by cytoplasmic vesicles. Protoplasma 100, 267–288.

Sikes, C. S., and Wilbur, K. M. (1982). Functions of coccolith formation. Limnol. Oceanogr. 27, 18–26.

Sikes, C. S., Roer, R. D., and Wilbur, K. M. (1980). Photosynthesis and coccolith formation: Inorganic carbon sources and net inorganic reaction of deposition. Limnol. Oceanogr. 25, 248–261.

Simkiss, K. (1986). The process of biomineralization in lower plants and animals—An overview. Syst. Assoc. Spec. Vol. 30, 19–38.

Simonsen, R. (1987). "Atlas and Catalogue of the Diatom Types of Friedrich Hustedt," Vol. 2. Cramer, Berlin. Copyright Gebrüder Borntraeger.

Towe, K. M. (1972). Invertebrate shell structure and the organic matrix concept. Biomineralization 4, 1–14.

Towe, K. M., and Cifelli, R. (1967). Wall ultrastructure in the calcareous foraminifera: Crystallographic aspects and a model for calcification. J. Palaeontol. 41, 742–762.

van der Wal, P., de Jong, E. W., Westbroek, P. de Bruijn, W. C., and Mulder-Stapel, A. A. (1983a). Polysaccharide localization, coccolith formation, and Golgi dynamics in the coccolithophorid Hymenomonas carterae. J. Ultrastruc. Res. 85, 139–158.

van der Wal, P., de Jong, E. W., Westbroek, P., de Bruijn, W. C., and Mulder-Stapel, A. A. (1983b). Ultrastructural polysaccharide localization in calcifying and naked cells of the coccolithophorid. Emiliania huxleyi. Protoplasma 118, 157–168.

van der Wal, P., de Vrind-de Jong, E. W., de Vrind, J. P. M., and Borman, A. H. (1984). Hymonomonas carterae—Calcification rate related to condition of coccosphere? GUA Pap. Geol. 20, 81–88.

van der Wal, P., de Bruijn, W. C., and Westbroek, P. (1985). Cytochemical and X-ray microanalysis studies of intracellular calcium pools in scale bearing cells of the coccolithophorid Emiliania huxleyi. Protoplasma 124, 1–9.

Watabe, N. (1967). Crystallographic analysis of the coccolith of Coccolithus huxleyi. Calcif. Tissue Res. 1, 114–121.

Watabe, N., and Wilbur, K. M. (1966). Effects of temperature on growth, calcification and coccolith formation in *Coccolithus huxleyi* (Coccolithinae). *Limnol. Oceanogr.* **11**, 567–575.

Westbroek, P., de Jong E. W., van der Wal, P., Borman, A. H., de Vrind, J. P. M., Kok, D., de Bruijn, W. C., and Parker, S. B. (1984). Mechanisms of calcification in the marine alga *Emiliania huxleyi*. *Philos. Trans. R. Soc. London, Ser. B* **304**, 435–444.

Westbroek, P., van der Wal, P., van Emburg, P. R., de Vrind-de Jong, E. W., and de Bruijn, W. C. (1986). Calcification in the coccolithophorids *Emiliania huxleyi* and *Pleurochrysis carterae*. I. Ultrastructural aspects. *Syst. Assoc. Spec. Vol.* **30**, 189–202.

Wilbur, K. M., and Watabe, N. (1963). Experimental studies on calcification of the alga *Coccolithus huxleyi*. *Ann. N.Y. Acad. Sci.* **109**, 82–112.

Williams, R. J. P. (1981). Natural selection of the chemical elements. *Proc. R. Soc. London, Ser. B* **213**, 361–397.

7 | Plant Mineralization— Photosynthesis and Cell Walls

The calcareous algae hold an important place in the biomineralization of marine environments and they are also present in freshwaters and soils. In number there are more than 100 calcifying genera and most algal divisions have representatives that deposit calcium carbonate (Table 7.1).

The activities of calcareous algae are most clearly seen in the formation of reefs and atolls (Table 7.2). We commonly assume that the mass of coral atolls is primarily the result of the growth of corals and red algae, but borings taken from the floor of several Pacific lagoons have shown that calcified segments of the green alga *Halimeda* may comprise the major portion of the sediment, emphasizing the importance of this particular genus in atoll formation (Hillis-Colinvaux, 1980). Encrusting red algae of the family Corallinaceae also have a special, important function in cementing and consolidating tropical reefs with a calcareous solid covering. In this way they stabilize the deposits produced by corals and other invertebrates against wave action. Marine blue-green algae may form mats which not only deposit calcium carbonate but also entrap calcareous sedimentary particles. Some algae may, however, have the opposite effect. Several individual species have the capacity for boring into limestone and so hastening its breakdown. In all cases, however, the algae are important photosynthesizers adding metabolites to food chains. The oceanic coccolithophorids make a major contribution in this way and encrusting reef algae provide a food source for numerous invertebrate and fish grazers.

Table 7.1 The Taxonomy of Calcareous Algae and Their Carbonate Species[a]

Algae	Polymorph of $CaCO_2$ deposited	Site of deposition	Habitat
Cyanophyta	Calcite (usually)	Extracellular in mucilage	Freshwater and marine
Cyanophyceae			
Rhodophyta			
Rhodophyceae			
Nemalionales	Aragonite	Extracellular in intercellular space	Marine
Cryptonemiales			
Peysonelliaceae	Aragonite	In cell wall (?)	Marine
Corallinaceae	Calcite	In cell wall	
Dinophyta			
Dinophyceae			
Peridiniales	Calcite[b]	In cell wall	Marine
Chrysophyta			
Haptophyceae	Calcite	In the Golgi[c]	Marine and freshwater
	Calcite (?)	In external mucilage	
Phaeophyta			
Phaeophyceae			
Dictyotaceae (*Padina*)[d]	Aragonite	Extracellular	Marine
Charophyta			
Charophyceae			
Charales (internodal cells)	Calcite	Extracellular	Freshwater
Charales (oogonia)	Calcite	Extracellular and in cell wall	
Chlorophyta			
Chlorophyceae			
Dasycladales	Aragonite	Extracellular in intercellular space	Marine
Caulerpales	Aragonite	Extracellular in intercellular space	Marine
Derbesiales (*Pedobesia*)[d]	Aragonite	In cell wall[e]	Marine
Conjugatophyceae			
Desmidiales (*Oocardium*)[d]	Calcite	Extracellular in mucilage	Freshwater (soils)

[a] Hillis-Colinvaux (1980), modified from Borowitzka (1977).
[b] Calcite is deposited only in the resting cysts.
[c] The coccoliths are formed within the Golgi and then excreted.
[d] The generic name in parentheses indicates that this is the only genus reported to be calcareous.
[e] Only in the disk-like stage of the life cycle.

Table 7.2 Calcification on Seaward Reef Flats Showing Concentrations of Corals
and Algae[a]

Locality	Seaward Reef Flats	kg $CaCO_{-3}$ $m^{-2}y^{-1}$
One Tree Reef[b]	Algal pavement (no coral)	4.0
One Tree Reef[b]	Coral reef flat zone	4.5
Lizard Island[c]	Seaward slope coral pinnacle	3.7
Lizard Island[c]	Reef flat (algal pavement)	3.8
Lizard Island[c]	Reef flat (coral/algal zones)	3.6
Eniwetok Atoll[d]	Reef flat (coral algal)	4.0
Eniwetok Atoll[d]	Reef flat (algal)	4.0

[a] From Kinsey and Davies (1979).
[b] Kinsey (1977).
[c] LIMER (1976).
[d] Smith (1973).

The extensive literature on calcareous algae has been summarized in reviews covering both morphological and physiological aspects by Darley (1974), Borowitzka (1982a,b, 1984; 1987), Klaveness and Paasche (1979), Hillis-Colinvaux (1980), Johansen (1981), and Pentecost (1985). In discussing calcareous algae we shall draw extensively from these sources.

I. Mineralizing Algal Systems

The morphology of algal mineralization takes a variety of forms depending on the taxon involved (Pentecost, 1985), but the forms can be classified as two basic types, extracellular and intracellular. Table 7.3 lists variations within these types (Borowitzka, 1982a).

In Borowitzka's classification, *extracellular mineralization* involves crystals forming on the surface of the outer cell wall and in the cell wall itself. The crystals on the outer surfaces do not exhibit any uniform orientation, indicating that they are not growing on or within oriented organic molecules. Crystals growing *within* cell walls are, however, frequently oriented as might be expected in this organized organic structure. If the outer wall is covered by a sheath, this also may be involved with crystals which either develop within the sheath or between the wall and the sheath (Böhm *et al.*, 1978). *Intercellular mineralization,* as the term implies, is associated with spaces enclosed between and bounded by cells; it is therefore properly considered extracellular mineralization. Intracellular mineralization is of common occurrence in invertebrates and higher

Table 7.3 Sites and Forms of Algal Mineralization[a]

Sites	Forms	Taxa
Extracellular		
Cell wall surface	Bands of calcite crystals on cell walls of internodal cells	Characeae (stoneworts) *Chara, Nitella*
	Concentric bands of fine aragonite needles on cell surface	Dictyotaceae (brown algae) *Padina*
	Surface encrustation of calcite crystals	Cladophoraceae (green algae) *Chaetomorpha*
		Oedogoniaceae *Oedogonium*
Intercellular	Fine aragonite needles in intercellular spaces in utricles	Udoteaceae (green algae) *Halimeda, Udotea*
		Dasycladaceae (green algae) *Neomeris*
	Intercellular crystals of aragonite and/or calcite which may form small bundles	Helminthocladiaceae (red algae) *Liagora*
		Chaetangiaceae (red algae) *Galaxaura*
Sheath	Bundles of aragonite needles in external sheath	Udoteaceae (green algae) *Penicillus, Udotea*
	Irregular bundles of needle-like crystals, usually aragonite	Scytonemetaceae (blue-green algae) *Plectonema*
	Granular to massive rhombohedral calcite crystals	Rivulariaceae (blue-green algae) *Rivularia*
Within cell walls	Calcite crystals, often clearly oriented	Corallinaceae (red algae) *Lithophyllum, Lithothamnium*
Intracellular	Calcified plates (coccoliths) of various forms, usually calcite. Formed within Golgi vesicles	Prymnesiophyceae (golden algae) *Emiliania, Cricosphaera*
Intra- and Extracellular	Oogonial cells, calcite crystals. Deposition begins in cell walls and fills cells with crystals	Characeae (stoneworts) *Chara, Nitellopsis*

[a] After Borowitzka (1982a).

plants. However, the only known algal group utilizing this method of calcium carbonate deposition is the Coccolithophoridae (Prymnesiophyceae) to which we have already referred (Chapter 6).

II. Mineralization and Cell Walls

The green (Chlorophyta) and red (Rhodophyta) marine algae deposit extracellular mineral. This general type of mineralization takes several forms: within the cell wall, as an encrustation on the wall or a cell filament (e.g., *Chaetopleura cornudamae*), or in proximity to the wall (as in *Liagora distensa*) (Table 7.3). In some algae there is an extra layer or sheath around the cell wall of the filaments; aragonite is often deposited within it, sometimes in discrete bundles of aragonite needles parallel to the cell wall, as in *Udotea conglutiana, Rhipocephalus phoenix,* and *Penicillus pyriformis.* It is assumed that the sheath plays some part in the regulation and orientation of these crystals. Finally, appressed cell filaments on the surface of a thallus may form an intercellular space in which calcium carbonate crystals are deposited, as in *Halimeda.* In each of these cases it is apparent that the site of calcification represents a microenvironment in which the formation of calcium carbonate is facilitated by being at least partially isolated from the surrounding seawater. Thus, the removal of carbon dioxide of protons from these regions will tend to initiate mineral deposition by increasing the pH.

   In discussing extracellular mineralization in these algae, we shall give brief accounts of their mineral deposition and the relationship of photosynthesis to their rates of calcification (see Pentecost, 1985).

A. Chlorophyta

These mineralizing green algae are represented in the families Udoteaceae (e.g., *Udotea, Halimeda*), Dasycladaceae (e.g., *Neomeris*), and Cladophoraceae (e.g., *Chaetomorpha*) (Table 7.3). Experimental studies of mineralization in Chlorophyceae have been carried out principally on *Halimeda* (Borowitzka, 1982a,b, 1984) and we shall limit our discussion to this genus.

1. The Mineralizing System.   *Halimeda* forms a single or branched series of segments each consisting of coenocytic filaments. One or more lengthwise filaments form the medulla of the segment, and their branches constitute the segment cortex (Fig. 7.1). The branches, which

Figure 7.1 Partial longitudinal section through a central portion of a mature segment of *Halimeda monile* showing the organization of coenocytic filaments. The horizontally oriented filaments are the medulla, the vertical filaments the cortex; the swollen structures of the cortex are the utricles; the peripheral ones, cone-shaped, adhere at their outermost edges to form a continuous and closed peripheral surface. In a complete longitudinal section of the sample, roughly three times the number of medullary filaments would be present and the cortex would extend below as well as above. Crystals of calcium carbonate form in the spaces between utricles and medullary filaments (Wilbur *et al.*, 1969).

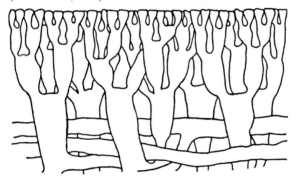

are usually somewhat swollen, are called utricles. Their outermost walls become pressed together and fused, forming a barrier which prevents the flow of seawater into the spaces between the utricles and the filaments. It is assumed that material can slowly diffuse in and out of this region, but it is within these interutricular spaces that mineral deposition takes place (Wilbur *et al.*, 1969; Borowitzka and Larkum, 1976b).

The mineral deposited by the Chlorophyta is aragonite. In *Halimeda monile* it is formed as long needles (see Fig. 7.2 of an unknown species of *Halimeda*) whereas in *Halimeda incrassata* the crystals are of granular or polygonal form. The crystals of both species are without preferred orientation. They first appear as small granules within the fibrous material on the wall of the filaments, which provides nucleation sites. As the crystals increase in size and number, they completely fill the interutricular spaces (Wilbur *et al.*, 1969).

2. Mechanisms of Mineralization. Mineral ions pass from the external seawater to the sites of mineralization (Fig. 7.3): either by diffusing through the outer appressed utricle walls or by transport through the coenocyte of the utricles. Since the area of the wall facing the interutricu-

Figure 7.2 Scanning electron micrograph of the extracellularly mineralizing calcareous alga *Halimeda* (species unknown) showing the thin, unoriented crystals of aragonite. A portion of the cell wall is visible in the upper right-hand corner. ×2880 (Weiner, 1986). (Reprinted with permission from *Critical Reviews in Biochemistry* **20**, 365–408.)

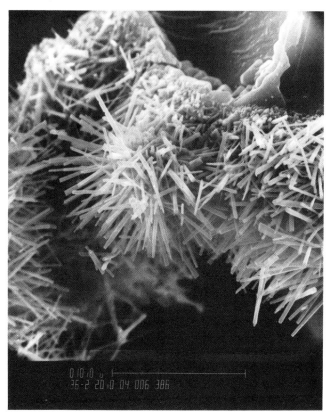

lar spaces is roughly 70% of the total wall area of the utricles (Borowitzka and Larkum, 1976a), most of the transported Ca^{2+} would move into the fluid within the interutricular spaces. Bicarbonate ions could presumably also enter this space by the same routes.

The rate of uptake of calcium from the seawater can be followed with [45]Ca and shows three phases (Fig. 7.4). There is an initial rapid accumulation followed by a slower phase, both of which are probably the result of [45]Ca binding to the walls of the utricles and exchanging with calcium on previously deposited crystals. The third phase with a relatively con-

Figure 7.3   Schematic representation of the postulated ion fluxes affecting $CaCO_3$ precipitation in *Halimeda*. The black dot at the plasmalemma indicates that this flux is postulated to be active. Passage of ions from the seawater to the intercellular space is by diffusion across 20 μm or more of cell wall of the appressed external cells (utricles); CO_2 for photosynthesis enters the cell by diffusion and CO_2 produced during respiration diffuses out of the cell; bicarbonate enters the cell probably by an energy-requiring process. Uptake of CO_2 from the intercellular space leads to a rise in pH and $[CO_3^{2-}]$, resulting in $CaCO_3$ (aragonite) precipitation; respiratory CO_2 evolution has the opposite effect. A light-stimulated proton efflux is also postulated. (From Borowitzka, 1983.)

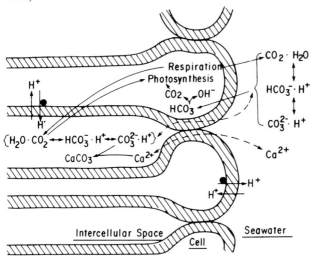

stant rate represents the accretion of new mineral. It is apparent from the results in Fig. 7.4 that the uptake of ^{45}Ca is enhanced by light and the experiment can be extended by using $H^{14}CO_3$ instead of ^{45}Ca. In this case, the rate of mineralization can be followed with greater precision and the fixing of carbon dioxide by photosynthesis can be measured simultaneously.

B. Rhodophyta

Four families of red algae have genera that calcify: the Corallinaceae, the noncorallinaceous Helminthocladiaceae (e.g., *Liagora*), and Chaetangiaceae (e.g., *Galaxaura*) form aragonite and/or calcite, while the coral-

Figure 7.4 Time course of ^{45}Ca uptake by plants of *Halimeda tuna* in the light (○), in the dark (●), and the dead plants (◆). Vertical bars are standard error calculated from at least 10 segments per data point. Plants were washed in unlabeled, ice-cold APSW for 1 min after each experimental period (Borowitzka and Larkum, 1976a).

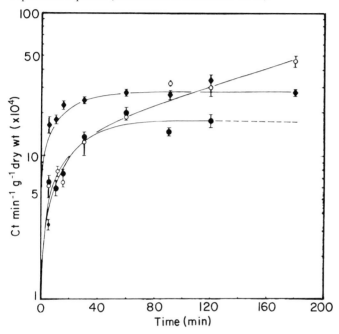

line Peyssonneliaceae (e.g., *Peyssonnelia*) produce calcite with a high Mg content (Table 7.1).

1. The Mineralizing System.   The Rhodophyta (Fig. 7.5), in contrast to the Chlorophyta, are primarily calcifiers of cell walls, although some do deposit mineral within intercellular spaces as well. Calcification of the walls is so extensive that the cells become encased except for the primary pit connections at the end (Fig. 7.6). The crystals deposited in the inner wall regions are usually oriented with the *c*-axis at right angles to the plasmalemma, although in some species the crystals appear to have a concentric orientation. Sections of the wall show an intimate relationship between the crystals and organic matrix (Borowitzka, 1982a). The fact that crystals are oriented within the wall strongly indicates that the crystals develop among oriented matrix molecules that influence their growth. One might expect that, if isolated walls could be gently decalci-

Figure 7.5 Articulated coralline algae. (A) *Metagoniolithon radiatum*; scale = 2 cm (Ducker, 1979). (B) *Cheilosporum proliferum*, a species having intergenicula with variously expressed lobes; scale in millimeters. (C) *Amphiroa anceps*, in which the intergenicula are flat; scale in millimeters (Johansen, 1969). [Reprinted with permission from "Coralline Algae: A First Synthesis" (H. W. Johansen, ed.). Copyright © 1981 CRC Press, Inc., Boca Raton, Florida.]

Figure 7.6 Scanning electron micrograph of a fractured thallus of *Lithothamnium australe* showing the calcite-impregnated cell walls. This alga has especially large calcite crystals which can clearly be seen to be oriented at right angles to the cell. [Courtesy of Borowitzka (1982b) from Townsend.]

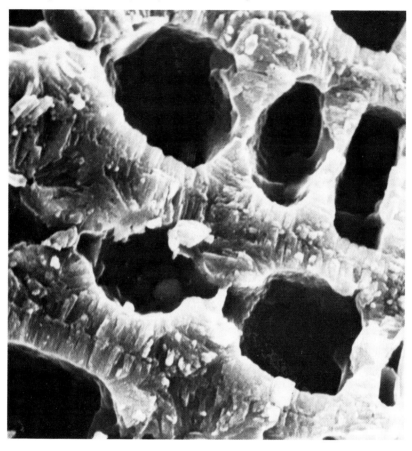

fied, then recalcification in a supersaturated solution might well result in the formation of oriented crystals approximating those normally present.

2. Mechanisms of Mineralization. Calcification in red algae can be considered as a variation of the mechanism present in green algae. In both, intercellular spaces are bordered by cells that apparently actively transport calcium ions (Miyata *et al.*, 1980; Okazaki *et al.*, 1982, 1984). A Ca-ATPase enzyme is certainly present in calcareous red algae but Okazaki (1977) found that it was by no means a universal feature of all algae. There may, therefore, be other ways of transporting calcium but, presumably, once the concentrations increase sufficiently within the walls and within the intercellular spaces, mineral deposition takes place. It would appear that crystal formation is inhibited in the walls of the green algae since crystals form only in the intercellular spaces of these species (see Borowitzka, 1984).

The presence of Ca-binding ligands in the organic matrix of mineralized structures has led to considerable speculation as to whether the cell wall can act as a nucleation site for mineral deposition. The binding pattern of Ca^{2+} in *Amphiroa folicea* suggests at least two binding sites in the cell wall, possibly COO^- of cellulose and SO_4^- of sulfated polysaccharides (Borowitzka, 1984). It should be pointed out, however, that crystal nucleation is not necessarily dependent on specific calcium binding sites and may occur simply by the provision of a solid surface.

We mentioned that some genera of Rhodophyta are calcitic and that others are aragonitic (Table 7.1). Okazaki *et al.* (1979) have approached the problem of the determination of the isomorph by adding acid-insoluble residues from a calcitic species and an aragonitic species to a solution from which calcite is normally precipitated. When an extract from the calcitic *Serraticardia maxima* was added to the solution, calcite and vaterite were precipitated, whereas the aragonitic *Galaxaura fastigiata* and *Galaxaura falcata* induced the formation of aragonite. Although these findings are suggestive, it is difficult to interpret them in terms of precise mechanisms.

III. Effects of Photosynthesis

It has long been recognized that there is a correlation between photosynthesis and the calcification rate of algae. A number of hypotheses have been advanced to explain this association but the suggestion that it is due to the removal of carbon dioxide was both the first hypothesis and the one for which there is most support. It proposed that the fixation of carbon dioxide leads to an increase in pH, i.e., $HCO_3^- \rightarrow CO_2 + OH^-$, which in turn facilitates the formation of calcium carbonate.

To produce this result, the pH change need not be large and the effect is, of course, favored by a small volume of fluid such as that enclosed within the intercellular compartment of an alga such as *Halimeda*. Borowitzka (1986) points out that the semi-isolation of such a microspace leads to a long diffusion path (see Fig. 7.3). The diffusion rate of carbon dioxide in water is roughly 1.5×10^{-10} cm^2 sec^{-1}, which is significantly greater than that of bicarbonate with a rate of 10^{-10} cm^2 sec^{-1}, so that photosynthesis will lead to a change in pH and an increase in carbonate ion concentration. In order for this system to be effective, the rate of carbon dioxide fixation must clearly exceed the rate at which it can enter the intercellular space from the outer medium and from metabolism. Borowitzka and Larkum (1976b) estimated that to accomplish this in *Halimeda tuna* the photosynthetic rate must exceed 100–200 mmol min^{-1} g dry wt^{-1}. The effect will be enhanced if there are further diffusion barriers to the entry of carbon dioxide from the external medium, and the mucilage between some algal cells may serve this purpose. In addition to *Halimeda*, genera with extracellular deposition sites separated from seawater by long diffusion paths include *Udotea, Penicillus, Cymopolia, Neomeris,* and the red algae *Liagora, Galaxaura,* and *Peyssonnelia* (Borowitzka, 1987).

Pulse–chase experiments in which algae are transiently exposed to NaH^{14}CO$_3$ and subsequently analyzed for organic and inorganic ^{14}C show that there is some recycling of carbon dioxide. Carbon dioxide released by the respiration of photosynthetic products may be trapped in the intercellular spaces and subsequently incorporated into aragonite crystals. These are revealing experimental findings, but it appears that once a certain level of photosynthesis is achieved there is a roughly linear relationship between the rate of photosynthesis and the rate of calcification in the apical segments of *Corallina officianalis* (Pentecost, 1978; Borowitzka, 1987).

The relation of photosynthesis to mineral deposition in *H. tuna* has also been studied by Borowitzka and Larkum (1976b). They compared the quantity of carbon that was fixed into organic material with the amount that was trapped as calcium carbonate at varying concentrations of total carbon dioxide (i.e., $[CO_2] + [HCO_3^-] + [CO_3^{2-}]$). The two curves were similar (Fig. 7.7) except that calcification reached saturation at a higher total carbon dioxide level. Clearly factors other than photosynthesis limit calcification. This appears to be borne out by two other findings. First, mineral deposition continued in the dark at about half the rate in light (see also Borowitzka, 1983). Second, a similar effect to darkness was produced by DCMU, an inhibitor of photosynthesis (Borowitzka and Larkum, 1976c). The reduction in calcium carbonate deposition in darkness and with DCMU was interpreted as resulting from the failure of CO$_2$ fixation. Mineralization under these conditions may also

Figure 7.7 The effect of increased CO_2 on gross photosynthetic O_2 evolution ($\square$), respiratory O_2 uptake in the dark ($\blacksquare$) ^{14}C fixation into the organic material (photosynthesis) ($\bigcirc$), and ^{14}C fixation into the inorganic material (calcification) ($\triangle$) in the light in *Halimeda tuna*. The pH was constant at pH 8·0. Note also the logarithmic scale (Borowitzka and Larkum, 1976b).

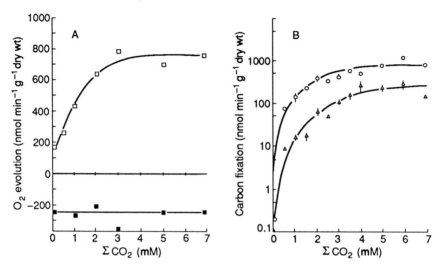

be dependent on the pumping of calcium ions into the interutricular fluid and this activity may also be reduced in the absence of ATP from photosynthesis.

The involvement of such energy sources is supported by the observation that *Gloeotaenium*, can only be made to calcify in the dark by adding metabolites such as succinate or acetate to the medium (Devi Prasad and Chowdary, 1981).

The calcification of green and red algae illustrates very simply four of the major components of biomineralization: an enclosed space, an increase in the activity of at least one ion, the possibility of nucleation by organic material, and the enhancement of calcification by photosynthesis. This simplicity is, no doubt, deceptive. There are many algal species that photosynthesize rapidly yet do not calcify. Conversely, algal mineralization can take place in the absence of photosynthesis. Perhaps the final lesson to be drawn from these studies is that no single metabolic activity results in biomineralization but that it does occur from the association of a few simple properties.

IV. Mineral Deposition in Freshwater Algae and Plants—Extracellular Mineralization

When calcium carbonate deposits occur in freshwater algae and plants, they are usually on the outer cell wall surfaces. The reasons for this have been revealed by some spectacular experiments.

A. Charophyceae

Calcification in freshwater green algae has been studied most extensively in the characean genera *Chara* and *Nitella*. Both form circular bands of mineral at intervals along the cylindrical internodal cells. The bands are formed of calcite crystals without preferred orientation (Borowitzka, 1982a). At the site of each mineral band, the pH is alkaline and may be as high as 10.5. The interband areas, however, lack mineral and are acid. The calcium carbonate of the mineralized bands is apparently precipitated from the external medium by the elevated pH which converts bicarbonate to carbonate ions.

The separation of alkaline and acid bands points immediately to corresponding local differences in ion transport across the plasmalemma. Spear (1969) examined this by separating calcified from noncalcified areas of internodal cells of *Nitella* by means of rubber membranes. The differences between the areas were correlated with the local extrusion of protons. More recent experimentation has been concerned with the movements of carbon dioxide, bicarbonate, hydrogen, and hydroxyl ions both inward and outward across the plasmalemma in relation to the separation of regions differing in pH.

Walker and Smith (1977) suggested that the alkaline bands were the result of proton influx into the cell, whereas Lucas (1985) considered that the alkalinity was caused by hydroxyl ion efflux. According to the hypothesis of Lucas (1985), the acid and alkaline bands were caused by a single process involving the assimilation of bicarbonate ions. The salient features of the model (Fig. 7.8) are as follows:

1. HCO_3^- enters the cell by cotransport with H^+.
2. An H^+-ATPase provides a supply of protons which produce the acid bands and drive the cotransport system.
3. The protons which enter the cell with HCO_3^- are recycled through the ATPase system. Some of the protons within the cell react with HCO_3^- to produce CO_2, which is fixed by photosynthesis.

Figure 7.8 Schematic representation of the plasmalemma-bound tranport complexes that may function to enable internodal cells of *Chara* to form alkaline and acid bands. $CaCO_3$ forms bands on a single cell (Lucas, 1985). (Reproduced with permission of American Society of Plant Physiologists.)

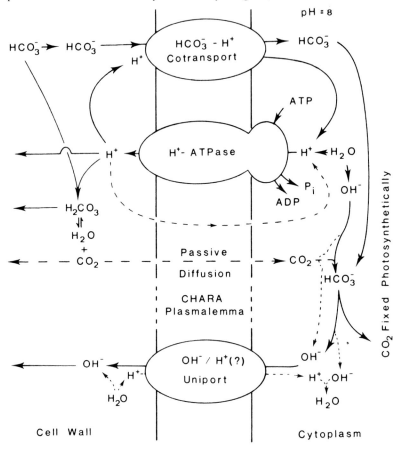

4. Protons in the outer medium react with HCO_3^- to produce CO_2, some of which enters the cell and is fixed.
5. The fixation of CO_2 causes OH^- to move out through the cell membrane, perhaps by a uniport system, producing the alkaline bands.

   An intriguing unsolved feature of the formation of these pH bands is the separation of the movements of individual ions into distinct areas of the membrane of a single cell.

   The role of photosynthesis in these processes has been neatly demonstrated by Okazaki *et al.* (personal communication) by embedding *Chara* in agarose gel containing bicarbonate ions, calcium ions, and phenol red. When exposed to light, alkaline bands appeared and calcium carbonate crystals were formed at the same regions. DCMU, an inhibitor of photosynthesis, reversibly inhibited the appearance of these bands.

B. Higher Plants

In higher freshwater plants, minerals are usually only deposited on the upper surface of the leaves, as in *Potomogeton, Elodea,* and *Myriophyllum.* The immediate reason for this was demonstrated by means of small pH electrodes which showed the upper mineralized surface to be alkaline whereas the lower unmineralized surface was acid (Prins *et al.*, 1980). In the presence of light, bicarbonate ions are thought to pass from the external medium across the lower surface accompanied by cations. A model indicating the possible pathways is shown in Fig. 7.9 (Prins *et al.*, 1982). At the upper surface hydroxyl ions are released and the resulting

Figure 7.9 Model for exogenous HCO_3^- utilization through the action of a proton pump. Two possible pathways are given for K^+ transport: in the left half through the cell wall and the right half via the symplast (Prins *et al.*, 1982).

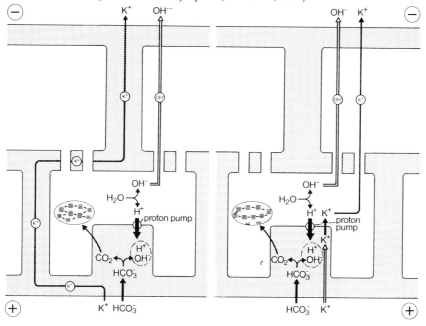

increase in pH assists in the formation of calcium carbonate, if sufficient calcium is present in the medium. The acidity at the lower leaf surface could result from an outwardly directed proton pump or from an outward diffusion of carbon dioxide. The change in acidity occurred in the absence of external bicarbonate, from which we can conclude that a proton pump is responsible. The pump, in addition to moving protons outward through the plasmalemma, could also provide carbon dioxide from the bicarbonate ions in the medium. The carbon dioxide could then diffuse into the cell and provide a carbon source for photosynthesis. In both the characean and the higher plant models, the hydroxyl ion for alkalinization at the outer cell surfaces is provided by the removal of carbon dioxide by photosynthesis.

The alkalinization and the formation of mineral on cell surfaces resulting from photosynthesis may have a major effect in the medium. In laboratory experiments on the water soldier (*Stratiodes aloides*), the plant and detached leaves can be shown to increase the pH and precipitate calcium carbonate in the medium. In profuse stands of this angiosperm in the shallow water of a lake, the amount of mineral deposited may be sufficient to produce large deposits of marl (Brammer and Wetzel, 1984).

V. Summary

1. Examples of extracellular, intercellular, and intracellular calcification occur in the algae.
2. Extracellular calcification occurs in both the red and green algae and involves the outer cell wall surfaces and intercellular spaces. The deposits may or may not be oriented.
3. There is evidence for the transport of calcium ions at some sites of calcification, although Ca^{2+}-ATPase is not universally present.
4. Photosynthesis enhances many forms of calcification. The most popular hypothesis to explain this suggests that carbon dioxide fixation induces carbonate ion formation.
5. Some freshwater algae and plants show pH changes on their surface associated with mineralization.
6. Marine algae show four of the main properties of calcifying systems, namely, (a) an enclosed space, (b) increased activity of at least one ion, (c) nucleation on a surface, and (d) enhanced activity associated with photosynthesis.

References

Böhm, E. L., Fütterer, D., and Kaminski, E. (1978). Algal calcification in some Codiaceae (Chlorophyceae): Ultrastructure and location of skeletal deposits. *J. Phycol.* **14**, 486–493.

Borowitzka, M. A. (1977). Algal calcification. *Oceanography and Marine Biology, Annual Review* **15**, 189–223.

Borowitzka, M. A. (1982a). Morphological and cytological aspects of algal calcification. *Int. Rev. Cytol.* **74**, 127–162.

Borowitzka, M. A. (1982b). Mechanisms in algal calcification. *Prog. Phycol. Res.* **1**, 137–178.

Borowitzka, M. A. (1983). Calcium carbonate deposition by reef algae: Morphological and physiological aspects. *In* "Perspectives on Coral Reefs" (D. J. Barnes, ed.), pp. 16–28. Clouston, Manuka, Australia.

Borowitzka, M. A. (1984). Calcification in aquatic plants. *Plant Cell Environ.* **7**, 457–466.

Borowitzka, M. A. (1986). Physiology and biochemistry of calcification in the Chlorophyceae. *In* "Biomineralization in Lower Plants and Animals" (B.S.C. Leadbeater, and R. Riding, eds.), pp. 107–124. Oxford Univ. Press, Oxford.

Borowitzka, M. A. (1987). Calcification in algae: Mechanisms and the role of metabolism. *CRC Crit. Rev. Plant Sci.* **6**, 1–45.

Borowitzka, M. A., and Larkum, A. W. D. (1976a). Calcification in the green alga Halimeda. II. The exchange of Ca^2 and the occurrence of age gradients in calcification and photosynthesis. *J. Exp. Bot.* **27**, 864–878.

Borowitzka, M. A., and Larkum, A. W. D. (1976b). Calcification in the green alga Halimeda. III. The sources of inorganic carbon for photosynthesis and calcification and a model of the mechanism of calcification. *J. Exp. Bot.* **27**, 879–893.

Borowitzka, M. A., and Larkum, A. W. D. (1976c). Calcification in the green alga Halimeda. IV. The action of metabolic inhibitors on photosynthesis and calcification. *J. Exp. Bot.* **27**, 894–907.

Brammer, E. S., and Wetzel, R. G. (1984). Uptake and release of K^+, Na^+, and Ca^{2+} by the water soldier. *Stratiotes aloides* L. *Aquat. Bot.* **19**, 119–130.

Darley, W. W. (1974). Silicification and calcification. *In* "Algal Physiology and Biochemistry" (W. P. D. Stewart, ed.), pp. 655–675. Univ. of California Press, Berkeley, California.

Devi Prasad, P. V., and Chowdary, Y. B. K. (1981). Effect of some organic sources on the dark calcification of the freshwater green alga Gloeotaenium. *New Phytol.* **87**, 297–300.

Ducker, S. C. (1979). The genus *Metagoniolithon* Weber–van Bosse (Corallinaceae, Rhodophyta). *Aust. J. Bot.* **27**, 67–101.

Hillis-Colinvaux, L. (1980). Ecology and taxonomy of *Halimeda*: Primary producer of coral reefs. *Adv. Mar. Biol.* **17**, 1–327.

Johansen, H. W. (1969). Patterns of genicular development in *Amphiroa* (Corallinaceae). *J. Phycol.* **5**, 118–123.

Johansen, H. W. (ed.) (1981). "Coralline Algae: A First Synthesis." CRC Press, Boca Raton, Florida.

Kinsey, D. W. (1977). Seasonality and zonation in coral reef productivity and calcification. *Proc. Third Internat. Symp. Coral Reefs* **2**, 383–387.

Kinsey, D. W. (1985). Metabolism, calcification and carbon production. I. Systems level studies. *Proc. Int. Coral Reef Congr., 5th* **4**, 505–526.

Kinsey, D. W., and Davies, P. J. (1979). Carbon turnover, calcification and growth in coral reefs. *In* "Biogeochemical Cycling of Mineral-Forming Elements" (P. A. Trudinger and D. J. Swaine, eds.), pp. 131–162. Elsevier, Amsterdam.

Klaveness, D., and Paasche, E. (1979). Physiology of coccolithophorids. *In* "Biochemistry and Physiology of Protozoa," (M. Levandowsky and S. H. Hutner, eds.), Vol. 1, 2nd Ed., pp. 191–213. Acadmic Press, New York.

LIMER (1976). Metabolic processes of coral reef communities at Lizard Island, Queensland.

LIMER 1975 Expedition team. *Search* **7,** 463–468.

Lucas, W. J. (1985). Bicarbonate utilization by *Chara*: A re-analysis. *In* "Inorganic Carbon Uptake by Aquatic Photsynthetic Organisms" (W. J. Lucas and J. A. Berry, eds.), pp. 229–254. American Society of Plant Physiologists, Rockville, Maryland.

Miyata, M., Okazaki, M., and Furuya, K. (1980). Initial calcification site of the calcareous red alga *Serraticardia maxima* (Yendo) Silva (Studies on the calcium carbonate deposition of algae III). *In* "The Mechanisms of Biomineralization in Animals and Plants" (M. Omori and N. Watabe, eds.), pp. 205–210. Tokai Univ. Press, Tokyo, Japan.

Okazaki, M. (1977). Some enzymatic properties of Ca^{2+} dependent adenosine triphosphatase from a calcareous red alga, *Serraticardia maxima* and its distribution in marine algae. *Bot. Mar.* **20,** 347–354.

Okazaki, M., Misonou, T., and Furuya, K. (1979). Particular Ca-binding substances in marine macro-algae. III. The influence of acid-insoluble residues in crystals from of $CaCO_3$ precipitated *in vitro*. *Bull. Tokyo Gakugei Univ. Ser.* 4 **31,** 215–221.

Okazaki, M., Furuya, K., Tsukayama, K., and Nisizawa, K. (1982). Isolation and identification of alginic acid from a calcareous red algae, *Serraticardia maxima*. *Bot. Mar.* **25,** 123–131.

Okazaki, M., Shiroto, M., and Furuya, K. (1984). Relationship between the location of polyuronides and calcification sites in the calcareous red algae *Serraticardia maxima* and *Lithothamnion japonica* (Rhodophyta, Corcellinacea). *Jpn. J. Phycol.* **32,** 364–372.

Pentecost, A. (1978). Calcification and photosynthesis in *Corallina affecionalis* L. using the $^{14}CO_2$ method. *Br. Phycol. J.* **13,** 383–390.

Pentecost, A. (1985). Photosynthetic plants as intermediary agents between environmental HCO_3^- and carbonate deposition. *In* "Inorganic Carbon Uptake by Aquatic Photosynthetic Organisms" (W. J. Lucas and J. A. Berry, eds.), pp. 459–480. Amer. Soc. Plant Physiol., Rockville, MD.

Prins, H. B. A., Snel, J. F. H., Helder, R. J., and Zanstra, P. E. (1980). Photosynthetic HCO_3^- utilization of OH^- excretion in aquatic angiosperms. Light induced pH changes at the leaf surface. *Plant Physiol.* **66,** 818–822.

Prins, H. B. A., Snel, J. F. H., Zanstra, P. E., and Helder, R. J. (1982). The mechanism of bicarbonate assimilation by the polar leaves of *Potamogeton* and *Elodes*. CO_2 concentrations at the leaf surface. *Plant Cell Environ.* **5,** 207–214.

Smith, S. V. (1973). Carbon dioxide dynamics: A record of organic carbon production, respiration, and calcification in the Enewetak windward reef flat community. *Limnol. Oceanogr.* **18,** 106–120.

Spear, D. G. (1969). Localization of hydrogen ion and chloride ion fluxes in *Nitella*. *J. Gen. Physiol.* **54,** 397–414.

Walker, N. A., and Smith, F. A. (1977). Circulating electric currents between acid and alkaline zones associated with HCO_3^- assimilation in *Chara. J. Exp. Bot.* **28,** 1190–1206.

Weiner, S. (1986). Organization of extracellularly mineralized tissues: A comparative study of biological crystal growth. *CRC Crit. Rev. Biochem.* **20,** 365–408.

Wilbur, K. M., Hillis-Colinvaux, L., and Watabe, N. (1969). Electron microscope study of calcification in the alga *Halimeda* (order Siphonales). *Phycologia* **8,** 27–35.

8

Plant Mineralization—Ions, Silicification, and the Transpiration Stream

Mineralization in plants occurs in a variety of genera (Table 8.1). The most widely distributed minerals are calcium oxalate, calcium carbonate and silica. In angiosperms and gymnosperms, calcium oxalate predominates while calcium carbonate is less frequently present (Arnott, 1973; Webb and Arnott, 1982). In marine algae, however, as in the invertebrates, calcium carbonate is the principal mineral. Silica, which is not crystalline, also has a very wide distribution. The minerals may take the form of descrete structures or they may simply impregnate existing cell walls. In vascular plants, mineral crystals are present in the leaves, flowers, stems, seeds, and roots. Part of this distribution appears to be related to the transpiration stream in which a flow of water containing silicic acid and ions passes from the roots and evaporates largely from the leaves. Another major site of mineral deposition is in the plant vacuole, which is not only a major storage site but is also involved in maintaining the turgidity of the cell. Clearly, mineralization in plants shows a number of differences from the systems found in animals, and ion and water fluxes are likely to have a considerable influence on the process. We will therefore describe those systems in some detail.

I. Ion Fluxes

The basic mechanisms of biomineralization in algae, fungi, and most higher plants can be introduced by considering ion movements in a single cell.

106

Table 8.1 Distribution of Crystalline or Amorphous Inorganic Salts in Plants[a]

Calcium salt	Angiosperms	Gymno-sperms	Pteridophyta, bryophyta and relatives	Algae	Fungi	Lichens
Oxalate	Found in most species. Not found in Cyperaceae and some members of Scrophulariaceae. Usually typical for each species, e.g., leaves, stems, roots, fruits, seeds, etc.	Found in most species	Found in fern leaves. Rarely found in Bryophyta	Reported occasionally	Crystalline deposits found on both vegetative and reproductive structures in many Fungi	Relatively common. Other oxalates (e.g., copper oxalate) also found
Carbonate	Found commonly in cell walls, heartwood, and occasionally in leaves. Aragonite in fruit in *Celtis*. Also as amorphous deposits in many families, e.g., Moraceae	Reported in cell walls	Found in cells of some species	Widely distributed	In hyphae, plasmodia, and fruiting bodies of many species, especially slime molds	
Sulfate	In pith of some plants		Reported in one or two subjects	Present in the Cyanophyceae		
Silicate	In association with silica in cystoliths		In Equisetales		Occasionally	
Citrate Tartrate Malate	Commonly found in fruits (e.g., pericarp), in seed epidermis, and in seed epidermis, and in some leaves				Products of fermentation of some yeasts	
Phosphate	Reported in leaves, stems, and heartwood of a few species					

[a] From H. J. Arnott, personal communication.

Table 8.2 Electrochemical Activity Ratios
Across Plasmalemma if the Ca^{2+} Activity in the
Cytoplasm is Taken as 0.002 μeg ml^{-1a}

Ca^{2+} in Medium Meq $Liter^{-1}$	$\bar{\mu}_j^o/\bar{\mu}_j^c$ (Plasmalemma)
20	25,900
2	12,100
0.2	2,400
0.02	386
0.002	30

[a] From Macklon and Sim (1981). If Ca^{2+} activity in the
cytoplasm is of the order indicated, the electrochemical
activity across the plasmalemma exceeds the flux ratio of
1.0 by a factor of 30 to 26,000, depending on the external
Ca^{2+} concentration. This indicates the expenditure of
considerable amounts of energy by a Ca^{2+} efflux pump in
the plasmalemma. (See original for further details.)

For our purposes, we shall simplify the structure of that cell and
consider that it has two main portions, the cytosol, or fluid phase of the
cytoplasm, and a vacuole. The cytoplasm is bounded externally by a
differentially permeable membrane, the plasma membrane or plasmalemma. The vacuole, which contains a solution differing in composition
from the cytosol, is also separated from the cytosol by a selectively
permeable membrane, the tonoplast (Fig. 8.1). Minerals formed by the
cell are deposited either external to the plasma membrane, in association
with the cell wall, or within the vacuole. Mineral deposits do not normally occur within the cytoplasm (Kausch and Horner, 1983). By means
of the proton microprobe low concentrations of mineral elements can be
measured in plant cells, and changes in concentration during development can be followed (Perry et al., 1984).

Calcium enters the cell across the plasmalemma by diffusion down an
electrochemical gradient. The gradient results from a low concentration
within the cytosol and is enhanced by a potential difference across the
plasmalemma (Fig. 8.1). By the use of the protein aequorin, which fluo-
resces in the presence of free calcium, it is possible to measure the free
calcium ion concentration in the cell. In the cytosol of the alga *Chara*,
Williamson and Ashley (1982) found a range of free calcium of 1 to
5×10^{-7} *M* with an average value of 2.2×10^{-7} *M*. Other studies of
calcium ion concentration in plant cells have indicated a range of 10^{-6}–
10^{-7} *M*. However, lower concentrations have been found in the cyto-

Figure 8.1 Working model of H$^+$-translocating ATPases and H$^+$-coupled solute transport systems in a simplified plant cell. The scheme is based on results from transport studies of intact cells and isolated vesicles of the tonoplast and plasma membrane. A plasma membrane and a tonoplast-type H$^+$-ATPase generate a proton electrochemical gradient which provides the driving force for active transport of inorganic and organic cations (C$^+$), anions (A$^-$), and sugars (S). Evidence for H$^+$-coupled transport using tonoplast vesicles includes antiport systems for H$^+$/Ca^{2+}, H$^+$/K$^+$, H$^+$/sucrose, and H$^+$/glucose, and potential-driven Cl$^-$ uptake. H$^+$-coupled transporters found on plasma membrane vesicles include a H$^+$/K$^+$ antiport and a H$^+$/IAA symport. Electrogenic, H$^+$-pumping ATPases may also exist on other subcellular membranes such as the ER and Golgi (Sze, 1985). (Reproduced with permission from the *Annual Review of Plant Physiology*, volume 36, © 1985 by Annual Reviews, Inc.

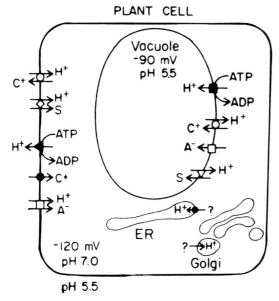

PLANT CELL

plasm of root hairs of *Lycopersicon esculentum* and *Brassica napus* using fura-2 fluorescence measurements (Clarkson *et al.*, 1988). These values would place the calcium ion concentration well below that of the external solutions and this situation is apparently maintained by an outwardly directed pump based on the enzyme Ca^{2+}-ATPase. There are, however, other mechanisms that are also involved in maintaining intracellular ion levels. The effect of changing the external calcium concentrations on the electrochemical activity across the plasmalemma of root segments of *Allium cepa* is shown in Table 8.2 (p. 108). As the calcium concentration is increased, membrane permeability may decrease (Macklon and Sim, 1981) with the result that the amount of calcium entering the cytosol may be less than expected.

From what we have said about the very low concentration of free calcium in the cytosol, it is immediately apparent that very little is likely to become mineralized unless ion activities are increased at some particular site. In fact, this is what appears to happen at the cell vacuole and there has been considerable interest in the types of calcium transport that occur in the tonoplast (Macklon, 1984) with evidence for both Ca/Mg-ATPase and Ca/H antiport systems (Chapter 4). It is clear, however, that any ion-transporting system of this type, especially if it involves a Ca/H antiport system, will favor the deposition of carbonate, phosphate, and oxalate minerals—all of which are deposited within plant vacuoles.

II. Cell Specialization

One of the striking aspects of the formation of calcium carbonate, calcium oxalate, and silica deposits in higher plants is its restriction to certain cells in the epidermis of the leaves. The phenomenon raises interesting questions about this cellular specificity. Two factors could be involved. One is the inability of tonoplasts to concentrate cations or anions in the vacuole to the level required for mineral deposition. Alternatively, there may normally be crystal inhibitors which are absent from mineralizing cells. Proteins, peptides, polyphosphate, magnesium, and citrate are all present in vacuoles (Goodwin and Mercer, 1982; Matile, 1978) and any of these may prevent mineral deposition. If substances inhibitory to crystallization were to enter only nonmineralizing vacuoles, then the permeability of the tonoplasts of the two cell types must be different. Such a difference might be examined by injecting labeled ions and organic compounds into the cytosol and observing their penetration into the vacuoles of mineralizing and nonmineralizing cells.

Alternatively, it may be that intracellular pH is the critical factor. Studies in recent years have emphasized that the distribution of solutes across the plasma membrane and the tonoplast is directly related to proton movement. Sze (1985), in a review of H^+-translocating ATPases, states: "We have learned that the primary active transport in higher plants is the electrogenic transport of H^+ extruded from the cell across the plasma membrane ... and probably pumped from the cytoplasm into the vacuole across the tonoplast...." This mechanism is considered the driving force for the movement of many cations, anions, and organic compounds and may play an important role in mineralization.

III. Isolated Vacuoles and Vesicles

An understanding of intracellular mineralization will require extensive data of three types: (1) the permeability of tonoplasts to mineral ions, (2) the transport mechanisms for mineral ions and related symport and antiport ions across the tonoplast membrane, and (3) organic and inorganic analyses of vacuole contents to indicate nucleating or inhibitor substances. These data are largely lacking at present. Ideally, in meeting these experimental needs, the vacuoles of mineralizing cells should be compared with vacuoles of nonmineralizing cells.

The study of cell vacuoles and vesicles has made considerable progress by the development of methods of differential centrifugation for the isolation of these organelles in quantities sufficient for enzyme analyses and transport experiments. Wagner (1972) and Wagner and Siegelman (1975), using this approach, demonstrated that large numbers of isolated vacuoles could be obtained from leaves, petals, and fruits. If the vacuoles contained pigment, as in *Tulipa* petals which have anthocyanins, the pigment was retained, showing that the tonoplast maintains its impermeability to the pigment. Since then, many studies have been carried out on the vacuoles and vesicles of various species (Lin *et al.*, 1977; Saunders and Conn, 1978; Boller, 1982; Thom and Komor, 1984; Mandala and Taiz, 1985; Sze, 1985). Such studies provide exciting opportunities for probing the molecular activities involved in biomineralization and should be applied to plant systems which form calcium oxalates, silica and calcium carbonate.

IV. Calcium Oxalate Deposition

A. Fungi

In 1887 de Bary wrote, "Calcium oxalate is a substance so generally found in the Fungi that it is quite unnecessary to enumerate instances of its occurrence." Having stated that, he went on to say where he had observed its absence: Peronosporeae, many Hyphomycetes, and species of *Bovista* and *Lycoperdon*. His illustrations show needlelike crystals on the hyphae of *Agaricus campestris*, irregular aggregates on the outer surface of *Phallus caninus*, and spheres of radiating needles in bladder like cells of the same species. Calcium oxalate in various mineral forms was well known to de Bary but he thought its occurrence in cells rather exceptional. Recent studies by Arnott (1982) and Arnott and Fryar (1984) have, however, described calcium oxalate in the hyphae of forest litter and compost and suggest that intracellular crystals may be more frequent than de Bary had supposed. Figure 8.2 illustrates the first emergence of needle-shaped crystals from the interior of hyphae and, at a later stage, shows large numbers of longer external crystals. Some of the crystals appear to originate within cell walls (Arnott and Webb, 1983).

Figure 8.2 Crystals emerging from a hypha. (a) Early stage. (b) later stage showing coiled young hypha; bars = 1 μm (Arnott and Fryar, 1984).

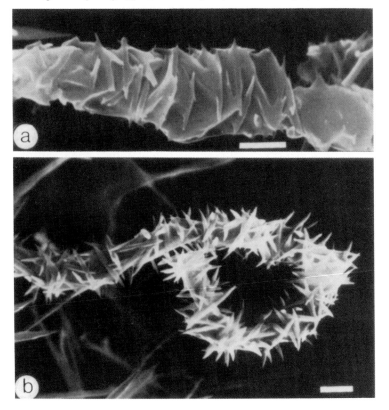

For a hyphal mat from compost, it is estimated that as many as 3×10^9 crystals with a total surface area of 450 cm^2 may occur per cm^2 of mat.

The relatively large amount of calcium oxalate formed by hyphae suggests that it is a metabolite that becomes enclosed in a membrane permeable to calcium. With a solubility product of Ksp = 1.78×10^{-9} crystals will form at low ion concentrations. The process could be a device for avoiding toxicity from either calcium and oxalate or it may be a way of deterring herbivores. Since crystals increase in size after they puncture the membrane from the inside, it appears that the hyphae cells still continue to metabolize oxalic acid under these circumstances. The mycorrhizal fungus *Pisolithus tinctorius* on the roots of *Pinus pinaster* also secretes oxalic acid, especially with alkaline pH and the addition of glucose (Callot *et al.*, 1985).

B. Lichens

Lichens deposit a number of minerals (Table 8.3). They resemble fungi in that they form calcium oxalates in two hydration states, the monohydrate (whewhellite) and the dihydrate (weddellite) (Jones *et al.*, 1980). The oxalic acid produced by lichens may be released from the organism and will react with minerals and cations in the immediate microenvironment. As a result, lichens may "weather" the surfaces of the rocks on which they grow by etching them, as can clearly be seen by scanning electron microscopy (SEM). Etching may then be followed by flaking and chipping of the surface. Similar results can be produced in the laboratory by adding oxalic acid to labradorite, a silicate mineral; by inducing the formation of weddellite crystals by the lichen myobiont in agar culture; or by culturing the common soil fungus, *Aspergillus niger*, with labradorite grains and observing etching of the grains (Jones *et al.*, 1980). Many lichens also produce a variety of weak phenolic acids that may also have a weathering effect but the major action can almost certainly be ascribed to oxalic acid. Along with the weathering of rock surfaces, metal cations including aluminum, iron, and magnesium may be extracted as well (Jones *et al.*, 1981).

The release of oxalic acid by lichens and the solubilizing of rock minerals may result in the extracellular formation of oxalate minerals. Examples are the formation of magnesium oxalate dihydrate associated with *Lecanora atra* (Wilson *et al.*, 1980) and manganese oxalate dihydrate associated with *Pertusaria corallina* (Wilson and Jones, 1984). Copper oxalate has also been reported (D. Jones and M. J. Wilson, unpublished results). Manganese oxalate and copper oxalate have not been previously re-

Table 8.3. Biomineralization Products Identified in Lichens[a]

Mineral Type	Composition	Name	Symmetry	Crystal Form
Oxalates	$CaC_2O_4 \cdot H_2O$	Whewellite	Monoclinic	Platey
	$CaC_2O_4 \cdot 2H_2O$	Weddellite	Tetragonal	Bipyramidal
	$MgC_2O_4 \cdot 2H_2O$	Glushinskite	Monoclinic	Distorted pyramidal
	$MnC_2O_4 \cdot 2H_2O$	Unnamed	Monoclinic	Blocky
	$CuC_2O_4 \cdot nH_2O$ ($n \cong 0.1$)	Unnamed	Orthorhombic	Platey and blocky
Iron oxides	$Fe_5HO_8 \cdot H_2O$	Ferrihydrite	Hexagonal	Gel-like
	$FeOOH$	Goethite	Orthorhombic	Very fine-grained
Clay minerals	$Al_2Si_2O_5(OH)_4$	Kaolinite	Monoclinic	Platey
	$Al_2Si_2O_5(OH)_4 \cdot 2H_2O$	Halloysite	Monoclinic	Fibrous
	$(M_y + nH_2)(Al_{2-y}Mg_y)Si_4O_{10}$	Montmorillonite	Monoclinic	Sheet-like
Hydroxycarbonate	$[Pb_3(CO_3)_2(OH)_2]$	Hydrocerussite	Hexagonal	Very fine-grained

[a] From Jones and Wilson (1986).

ported for organisms (Lowenstam and Weiner, 1983). These extracellular deposits probably have no organic matrix associated with the crystals.

C. Higher Plants

To say that calcium oxalate is widespread in higher plants is almost an understatement. McNair (1932) reported that calcium oxalate crystals had been observed in representatives of 215 families of gymnosperms and angiosperms. Chattaway (1955, 1958) indicated that the mineral was present in wood of more than 1000 genera of 180 families. The amount of oxalate in plants may be considerable, and values in excess of 10% of the dry weight are not uncommon (Zindler-Frank, 1976). Studies of the ultrastructure of the tissues and crystals, crystallography, and metabolic aspects have been reviewed by Arnott and Pautard (1970), Arnott (1973, 1976), Zindler-Frank (1976), and Franceschi and Horner (1979).

Oxalic acid, once synthesized by a plant, may occur as a free acid (Ranson, 1965) or as a soluble salt (Zindler-Frank, 1976) and is present within the cytosol. It will then move through the tonoplast into the vacuole. The mechanism by which oxalate enters the vacuole, whether by diffusion or through enzyme action, is not clear (Marty et al., 1980). Although ATPases are present in the tonoplast and might have a role in the transport of oxalate, their involvement is unknown. Once oxalate has crossed the tonoplast it can then react with calcium ions within the vacuole and crystallize as calcium oxalate. An active transport of calcium through the tonoplast appears quite likely (Schumaker and Sze, 1985). This, together with the low solubility product for calcium oxalate, would be favorable for mineral formation. However, the synthesis of oxalic acid by a species does not necessarily result in deposition of oxalate crystals in that idioblasts, the cells in which the crystals form, may not be present, and free calcium may not be available in sufficient concentration within the cell vacuole.

Since the solute composition within the vacuole may well vary among species and among cells of the same species it is not surprising that there are differences in the crystal habit of calcium oxalate. Whewhellite, the monohydrate form of calcium oxalate, may take many forms whereas weddellite has been identified only in prisms and druses.

In relation to the control of crystal form, Cody and Horner (1985) point out that, in dicotyledons, crystals of the same shape are generally located in the same position in all members of a single species and/or family. This uniformity suggests a genetic influence expressed at the site of crystal formation. In addition, the membrane of the enclosing vacuole or the presence of oriented molecules within the vacuole may help to determine the shape of the crystal. The individual crystals of the raphides of *Yucca* and *Spirodela* and the druses of *Helianthus* illustrate this possibility. The cross-sectional shape of the crystals is foreshadowed in the shape of the vacuoles prior to the development of crystals (Arnott, 1966).

Crystals, especially elongate ones, are frequently oriented within the cell. This is true, for example, of the single crystals of the jackbean *Canavalia ensiformis* (Frank and Jensen, 1970) and the raphides of the duckweed *Spirodela oligorrhiza* (Ledbetter and Porter, 1970). In both, the crystals are oriented parallel to the long axis of the cell. The orientation and proximity of individual crystals within a raphide are such that a closely packed bundle of parallel crystals is formed, each crystal within its vacuole. One can reasonably suppose that in these species specific molecules within the cytoplasm determine the orientation of the vacuoles and thus of the crystals themselves. The raphide bundles of adventitious roots of *Vanilla planifolia* present an interesting example of this

Figure 8.3 Raphide bundles of calcium oxalate in adventitious roots of *Vanilla planifolia*. Note that the bundles in adjacent idioblasts are perpendicular to each other. Nu, nucleolus; bar = 10 μm (Kausch and Horner, 1983).

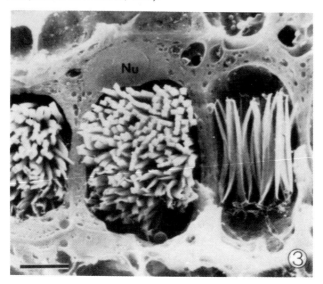

orientation (Kausch and Horner, 1983). The bundles can reorient to lie in a position perpendicular to the long direction of growth of the row of idioblasts, whilst in adjacent idioblasts, the bundles may be perpendicular to each other (Figure 8.3).

That calcium oxalate crystals form within a membrane-bound chamber of the vacuole has two features of considerable interest.

1. The chamber morphology is apparently achieved by self-assembly of molecules under genetic control. This governs the general shape of the crystals.
2. The completion of the chamber initiates crystal growth. Three mechanisms may make this possible: (a) The use of macromolecules to form the chamber walls may alter the membrane of the vacuole, resulting in the influx of Ca and oxalate ions. (b) Nucleation sites may become available on the interior of the chamber walls, resulting in crystal formation. (c) Chamber formation may remove inhibitory substances within the vacuole, so permitting nucleation on the interior chamber walls.

V. Silica Deposition in Higher Plants

At the outset of our discussion of silica, attention is called to the volume edited by Simpson and Volcani (1981) as a most useful source of information. Other helpful sources are reviews by Lewin and Reimann (1969) and Jones and Handreck (1967); a symposium volume on the biochemistry of silicon edited by Bendz and Lindquist (1978); and a review on transport by Raven (1983). In the sections that follow, we shall draw on all of these.

The distribution of silica is very wide, occurring in algae and higher plants (Table 8.4). Diatoms are a striking example, with concentrations of 95% silica in their walls. Of the higher plants, the grasses, cereals, horsetails, and sedges have relatively high concentrations (Sachs, 1887; Kaufman *et al.*, 1981). The dry matter of rice husks at harvest may contain 20% silica (Garrity *et al.*, 1984) and silica slag is regularly applied to rice paddies where silicification helps to keep the leaf blades erect and improves the yield. Concentrations in the Gramineae may be 10 to 20 times that in legumes and other cotyledons, and such differences are present even when these plants are grown in the same soil (Jones and Handreck, 1967). Silica is also common in broad-leaf trees and in conifers. Amos (1952, quoted in Scurfield *et al.*, 1973) lists 440 species of trees

Table 8.4 Distribution of Silica in Plants

Algae	Higher Plants[a]
Chrysophyceae (golden algae)	Equisetophyta
Tribophyceae (Xanthophyceae) (yellow-green algae)	Equisetaceae (horsetails)
Bacillariophyceae (diatoms)	Pteridophyta
Silicoflagellineae (flagellates with netlike skeletons)	Spermatophyta
	Poaceae (grasses)
	Cyperaceae (sedges)
	Commelinaceae (spiderworts)
	Zingiberaceae (ginger family)
	Cannabaceae (hemp family)
	Urticaceae (nettles)
	Fabaceae (pea family)

[a] From Kaufman et al. (1981).

in 144 genera and 32 families in which silica occurs in wood. Silica is present in both dicotyledons and monocotyledons, but the amounts deposited by monocotyledons are probably greater (Parry et al., 1984).

X-ray diffraction, nuclear magnetic resonance, infrared spectrum analyses, and electron microscopy have shown most silica in plants and diatoms to be amorphous rather than crystalline and to occur as a silica gel containing tightly bound water and many Si-OH units. Silicon in this form is called silica or opaline silica (Jones et al., 1966; Lewin and Reimann, 1969; Scurfield et al., 1974; Mann et al., 1983). The general formula can be written $[SiO_{n/2}(OH)_{4-n}]_m$ where n can be 0–4. The monomer is $Si(OH)_4$ (Mann et al., 1983). However, in *Fragaria chilensis* and *Equisetum hyemale* some crystalline silica is present along with amorphous silica (Sterling, 1967).

A. Sites and Forms of Silica Deposition

Silica deposits are of three general types: (1) cystoliths which are bodies of various forms and size within the lumina of cells; (2) impregnation of cell walls; and (3) extracellular deposits within spaces between cells (Scurfield et al., 1974; Jones and Handreck, 1969). These types are found in a variety of sites: leaves and shoots (Sachs, 1887; Jones and Hay, 1975; Kaufman et al., 1981), stems (Amos, 1952), and roots and rhizomes (Sangster and Parry, 1981). Examples of cystoliths within a cell lumen are shown in Figure 8.4. In the epidermal layer of leaves, silica is present in the wall layers (Sakai and Thom, 1979) or external to the cell wall

Figure 8.4 (a) Side view of isolated cystolith showing general morphology as seen in the SEM. Note the concentric bands seen at the broken surface of the stalk. (b) Side view of an isolated cystolith, which was demineralized using EDTA, as seen in the SEM. Note that the stalk and basal part of the cystolith are composed of dissimilar elements. The internal structure of the stalk was not revealed by EDTA treatment (Arnott, 1980).

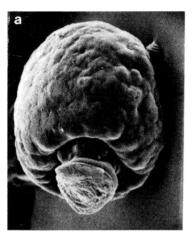
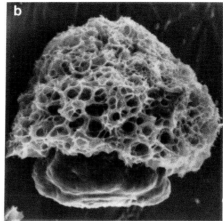

beneath the cuticle (Postek, 1981). Within a tissue, for example, in the epithelium of leaves and grasses (Figure 8.5), only certain cells show intracellular deposits. These cells are called idioblasts.

Identification of silicon in deposits *in situ* is commonly made in whole cells by energy-dispersive X-ray spectroscopy (EDX) (Kaufman *et al.*, 1969; Dinsdale *et al.*, 1979; Postek, 1981; see Kaufman *et al.*, 1981, for other methods). It should be realized that this method as applied to whole cells cannot differentiate between internal silica and silicon deposited in the cell wall. The method is useful, however, in determining the stage of development at which cells in the plant become silicified. Sakai and Sanford (1984) have shown by the EDX method that in sugar cane the accumulation of silicon proceeds at different rates in different cell types. Differences in silica accumulation in individual epidermal cells of a leaf are also readily made evident in secondary electron images (Sakai and Sanford, 1984).

The developmental sequence of silicification in the intercellular spaces of roots is illustrated in *Molinia caerulea* (Montgomery and Wynn Parry, 1979; Sangster and Parry, 1981). The earliest deposits are spheres of silica lining these spaces (Figure 8.6). The next stage corresponds with the formation of bricklike bodies that develop from the spheres. Deposi-

Figure 8.5   A scanning electron micrography of crenate-vertical silica bodies in the abaxial epidermis of *Olyra latifolia* Linnaeus. ×867. (Courtesy of P. G. Palmer.)

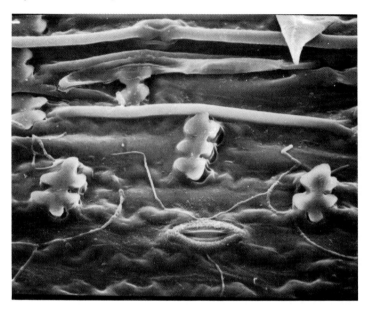

Figure 8.6   (a) Silica deposition in an intercellular space of the root of *Molinia caerulea* illustrating that the growth is centripetal. Spherical units, formed inside the original silica lining, are beginning to coalesce. Electron-dense fibrillar material remains in the central cavity. ×23,700. (b) A mature silica deposit completely fills the root intercellular space. Some of the electron-opaque fragments of this solid deposit are missing. no deposits are visible in the adjacent cell wall lamellae. ×17,100 (Sangster and Parry, 1981).

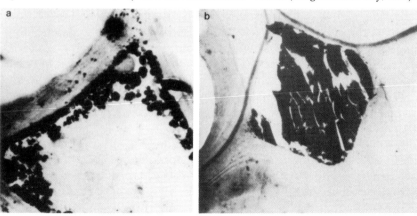

tion within the intercellular space then continues centrally with sphere formation, their coalescence, and finally the formation of solid masses occupying the cavity. In fact, the filling of the intercellular space may be so complete that it forms a precise cast of the region, including pits on the cell walls. Casts of intracellular lumina are also formed. By chemically removing the organic portions of plants, these siliceous casts can be collected for analysis.

B. Mechanisms of Silica Deposition

1. Movement of Monosilicic Acid. The siliceous deposits in higher plants are thought to be derived from the monosilicic acid [$Si(OH)_4$] of the soil. The concentration of monosilicic acid in the soil water may be increased by exudates from roots of plants of high silica content that depolymerize silicates and by soil humic acids that dissolve silicates (Sangster and Hodson, 1986). The monosilicic acid enters the root cortex, moves through the cortex to the endodermis, and then upward in the xylem vessels. The method of uptake from the soil varies according to the species and may be passive or active, while in some plants the silicic acid may actually be excluded from entering the root (Birchall, 1978). The tissue of major resistance in silicic acid uptake, as for other solutes, is almost certainly the endodermis (Parry et al., 1984).

The present view of the upward movement of monosilicic acid from the roots to the stems and leaves was stated by Sachs in his lectures published in 1887: "It is conveyed there with the water of nutrition, and, on evaporation of this, remains behind in the outer cell wall." Today, we would say that the upward movement is in the transpiration stream in the xylem. The monosilicic acid then passes outward through the bundle sheath and mesophyll to the epidermis (Sangster and Parry, 1981; Kaufman et al., 1981; Montgomery and Wynn Parry, 1979). En route, in the xylem transport vessels, some of the monosilicic acid moves outward and is deposited as polymerized silica within the roots in intercellular spaces, in cell walls, and in cell lumina (Parry et al., 1984). Within the leaves, the monosilicic acid passes to idioblast cells, guard cells, long cells, prickles, basal cells of microhairs, cell walls, and a layer immediately beneath the cuticle. During its movement to these final sites of deposition, the monosilicic acid must, of course, remain unpolymerized. The maintenance of the unpolymerized state of monosilicic acid during transport could perhaps be the result of association with organic compounds (Birchall, 1978; Iler, 1978; Weiss and Herzog, 1978; Williams, 1978) (Figure 8.7). Complexes may result from hydrogen bonding with

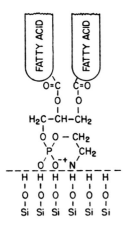

Figure 8.7 Association with organic compounds is a possible mechanism for preventing polymerization of monosilicic acid during transport.

O, NH_2, COOH, and other groups present in tissue fluids (Birchall, 1978). Whatever the mechanism for preventing polymerization, it is counteracted at deposition sites where polymerization takes place. For an understanding of silicification in higher plants, studies are clearly needed on the state of monosilicic acid during transport and the unpolymerized–polymerized transition at sites of deposition.

One way that polymerization from $Si(OH)_4$ is presumed to be accomplished is by an increase in concentration through evaporation of water as the $Si(OH)_4$ moves upward in the plant. This explanation has been advanced for silicification of vessels of wood fibers (Scurfield *et al.*, 1974). Evidence for the increase in concentration has been obtained by comparing the concentration of silica where the plant is growing to that in xylem sap. Such measurements confirm an increase in concentration for *Equisetum* (Kaufman *et al.*, 1981), and rice (Takahashi and Okuda, 1962, cited by Kaufman *et al.*, 1981). In addition, a number of studies have shown that water loss through transpiration can play a role in the silicon uptake of leaves (Jones and Handreck, 1969). In fact, the SiO_2 content of some plants, such as dryland Gramineae, corresponds to the uptake of water and the $Si(OH)_4$ concentration of the soil solution. A reduction in transpiration is commonly, although not invariably, associated with a reduction in the silica content of the leaves (Jones and Handreck, 1969). For example, in the heath grass *Sieglingia decumbens* deposition of silica does not take place if the leaves are coated with oil (Sangster and Parry, 1971).

If the movement of monosilicic acid from the soil to the deposition sites in the plant were entirely the result of loss of water through transpiration and diffusion of $Si(OH)_4$ down concentration gradients, then metabolic energy would not be required. Although there is substantial support for the passive movement of monosilicic acid in the experimental literature, a number of studies indicate that the passage of this compound through the roots may be more complex, at least in some plants. Lewin and Reimann (1969) and Raven (1983), in summaries of studies of various species, cite exceptions to passive movement. These include (1) entrance of monosilicic acid at a rate greater than that associated with transpiration and the concentration of $Si(OH)_4$ in the external solution; (2) monosilicic acid in xylem sap at concentrations several hundred times that in the external solution; (3) absorption of monosilicic acid independent of water absorption; and (4) effects of inhibitors suggesting active transport (see also Jones and Handreck, 1969; Sangster and Parry, 1971). Active transport of $Si(OH)_4$ is thought to occur in *Oryza, Hordeum,* and *Phaseolus* (see Raven, 1983). Although the transport can be inhibited, it is not clear whether the transport is active or mediated. An examination of root structure suggests that the endodermis is likely to be a site of active transport.

2. Initiation of Silica Deposition.   There are a number of conditions that may initiate or inhibit polymerization and the formation of siliceous structures in plants. As we have mentioned, one factor long considered important in polymerization is the loss of water through transpiration with a resulting increase in silicic acid concentration. This explanation has been recently considered in silica deposition in the macrohairs of the canary grass millet *Phalaris canariensis* (Sangster and Hodson, 1986; Hodson and Sangster, 1987). After emergence of the inflorescence, silica is first deposited in the outer cell wall layers and later throughout the wall. This could be taken to be the result of a passive increase in silica concentration resulting from transpiration, but it could also come about from an active outward transport of silica from the protoplast, as the authors suggest. It is probable that factors other than transpiration are involved. In the epidermis of leaves, only a few cells normally form phytoliths, implying some form of cellular specificity. Interestingly, cells that are not normally silicified may become so as they age, with the result that in the Gramineae the amount of silica increases as the leaves become older. The reason for the absence of silicification in the younger cells is unknown. Obvious possibilities are the impermeability of the cells to $Si(OH)_4$, an intracellular environment that inhibits polymerization, or the absence of nucleation centers. The interesting suggestion has been made that with aging and senescence the accumulation of cellular debris may provide these nucleation centers (Sangster, 1970b; Montgomery and Wynn Parry, 1979). Unfortunately no investigations have been carried out to determine whether injury to individual cells would induce silicification. The effects of cell organelles on the polymerization of monosilicic acid could presumably be examined *in vitro*. In a similar way, it should be possible to study the inhibition of polymerization by the cytosol of young cells.

Figure 8.8 (a) *Sieglingia decumbens*. A comparison of the absolute number of opal phytoliths in tiller bud leaves at the 16-day harvest for two silica concentrations. (b) Absolute number of opals in idioblasts of leaf 1 to 50 and 100 ppm silica after 2, 4, 8, and 16 days (Sangster, 1970a).

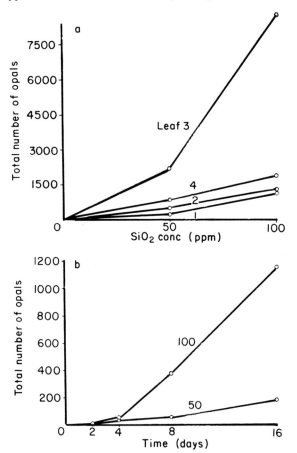

It has already been mentioned that the first morphological structures to appear in silicification were small rounded bodies contiguous with cell walls in intracellular spaces. These bodies are also found in tracheary cells of *Magnolia* (Postek, 1981) and in diatoms, amoebae, and radiolarians (Sangster and Parry, 1981). The association of these first bodies with the cell wall suggests that molecules on that surface might initiate polymerization, but it should be remembered that the bodies also form away from the walls as silicification develops centrally.

One way of studying the factors affecting intracellular silicification is to use the *in vitro* systems mentioned above. Such a method has been used with young leaves of the heath grass *Sieglingia decumbens* to follow the penetration of soluble silicon and its deposition (Sangster, 1970a,b; Sangster and Parry, 1971). Leaves were placed in nutrient solutions containing silica, removed at intervals over several days, and the number of phytoliths counted (Fig. 8.8). After 2 days, polymerized silica deposits were present.

During intracellular silicification, silica may react with proteins and cause their denaturation (Iler, 1978). Cationic and other polar groups of proteins form bonds with silica particles and, as a result, internal hydrogen bonds of the protein molecule may be stretched and broken. These reactions would seem to preclude the presence of silica bodies within *living* cells in that the vacuolar membrane and the plasma membrane could be altered and proteins of the cytosol be denatured. This, however, is an extreme view since molecules such as phospholipids may provide protective layers over the silica. Such bonding could occur by H bonds on the phosphate oxygen or by strong base cation–anion pairing see (Iler, 1978). The action of phospholipids on silica polymerization is not known and should be examined *in vitro* in relation to intracellular silicification.

As Williams (1978) has said ". . . silicon in life here on earth is little understood as yet. There is a lot of work to be done."

VI. Calcium Carbonate Deposition

The families of most importance in calcium carbonate deposition are the Urticaceae, Moraceae, Acanthaceae, and Cannabinaceae. In angiosperms the deposits occur within cells, in cell walls, and on the surface of the leaves of some freshwater plants. They are also present in epidermal cells of leaves, in heartwood, vessels of wood, and seeds. In gymnosperms calcium carbonate has been reported to be restricted to cell walls.

In fact, the major amount of calcium carbonate of plants occurs within or associated with cell walls. (Arnott and Pautard, 1970).

In the epidermis of leaves calcium carbonate is deposited together with silica in curious stalked structures called cystoliths (Fig. 8.4). The cystoliths of *Ficus elastica* (Moraceae) are formed from extensions of the cell wall which grow into the lumen of the cell. In *Beloperone californica* (Acanthaceae), Okazaki *et al.*, (1984) have described a somewhat different method of formation in which the cystolith originates in a vacuole attached to the cell wall. The rounded body portion of the cystolith appears to be the chief region of calcium carbonate deposition, whereas silica is in higher concentrations in the stalk (Arnott, 1980; Hiltz and Pobeguin, 1949). The deposition of both calcium carbonate and silica in different parts of a structure formed by a single cell is most unusual and may be unique in biomineralization. The mechanisms involved certainly deserve further experimental attention. The crystallography of cystoliths is not clear. The calcium carbonate appears in most cases to be amorphous, although small crystals may be present in *Morus*. Vaterite has been identified but its proportion has not been determined.

Decalcification of the cystoliths of *Morus* reveals a rather well-formed organic matrix (Arnott, 1980; Fig. 8.4). Histochemical tests of the matrix show the presence of cellulose, other neutral and acidic polysaccharides, and protein (Arnott, 1980; Okazaki *et al.*, 1984).

The physiological significance of cystoliths is obscure, but it appears possible that they may be involved in calcium or carbonate metabolism. In some species (e.g., *Morus niga*) they remain intact when the leaf is shed but in others (e.g., *F. elastica*) they are demineralized before leaf fall (Arnott, 1980).

VII. Summary

1. There are three systems of mineralization in the higher plants involving the deposition of calcium oxalates, silica, and calcium carbonates.
2. The formation of calcium oxalates is widespread but in the higher plants it is always intracellular and is often in specialized cells called idioblasts.
3. Silica enters plants as monosilicic acid and is deposited in phytoliths within the lumina of cells, as an impregnation of the cell walls, or as extracellular deposits between cells. Despite this, there is evidence for the control of mineral deposition at specific sites.
4. Calcium carbonate is deposited as an amorphous mineral in cystolith cells that occur mainly in leaves.

5. The mineralization processes in plants should be seen in relation to the unicellular activities of some algae (Chapter 6); the relationship of photosynthesis and cell walls (Chapter 7); and the deposition of particular minerals such as calcium oxalate, calcium carbonate, and silica (Chapter 8). There has clearly been a diversity of evolutionary strategies among these organisms, and mineral deposition can be related to functions as different as skeletal support and the deterrence of herbivores and encompass activities of vacuolar membranes and transpiration water loss.

6. A number of new techniques, such as the isolation of vacuoles and the use of tissue culture methods, could usefully be applied to problems such as the specificity of mineralization in certain cells and the initiation and control of silica polymerization in plants.

References

Amos, G. L. (1952). Silica in timbers. *CSIRO Aust. Bull.* **267**.

Arnott, H. J. (1966). Studies on calcification in plants. *In* "Third European Symposium of Calcified Tissues" (H. Fleisch, H. J. J. Blackwood, and M. O. Owens, eds.), pp. 152–157. Springer-Verlag, Berlin.

Arnott, H. J. (1973). Plant calcification. *In* "Biological Mineralization" (I. Zipkin, ed.), pp. 609–627. Wiley, New York.

Arnott, H. J. (1976). Calcification in higher plants. *In* "The Mechanisms of Mineralization in the Invertebrates and Plants" (N. Watabe and K. M. Wilbur, eds.), pp. 55–78. Univ. of South Carolina Press, Columbia, South Carolina.

Arnott, H. J. (1980). Carbonates in higher plants. *In* "The Mechanisms of Biomineralization in Animals and Plants" (M. Omori and N. Watabe, eds.), pp. 211–218. Tokai Univ. Press, Tokyo, Japan.

Arnott, H. J. (1982). Calcium oxalate (weddellite) crystals in forest litter. *Scanning Electron Microsc.* **3**, 1141–1149.

Arnott, H. J., and Fryar, A. (1984). Raphide-like fungal crystals from Arlington, Texas compost. *Scanning Electron Microscr.* **4**, 1745–1750.

Arnott, H. J., and Pautard, J. G. E. (1970). Calcification in plants. *In* "Biological Calcification: Cellular and Molecular Aspects" (H. Schraer, ed.), pp. 375–446. Appleton-Century-Crofts, New York.

Arnott, H. J., and Webb, M. A. (1983). The structure and formation of calcium oxalate crystal deposits on the hyphae of a wood rot fungus. *Scanning Electron Microsc.* **4**, 1747–1750.

Bendz, G., and Linquist, I., eds. (1978). "Biochemistry of Silicon and Related Problems." Plenum, New York.

Birchall, J. D. (1978). Silicon in the biosphere. *In* "New Trends in Bioinorganic Chemistry" (R. J. P. Williams and J. R. R. F. Da Silva, eds.), pp. 209–252. Academic Press, New York.

Boller, T. (1982). Enzymatic equipment of plant vacuoles. *Physiol. Veg.* **20**, 247–257.

Callot, G., Mousain, D., and Plassard, C. (1985). Concentrations du carbonate de calcium sur les parois des hyphes myceliens. *Agronomie* **5**, 143–150.

Chattaway, M. M. (1955). Crystals in woody tissues. Part 1. *Trop. Woods* **102,** 55–74.
Chattaway, M. M. (1956). Crystals in woody tissues. Part 2. *Trop. Woods* **104,** 100–124.
Clarkson, D. T., Brownlee, C., and Ayling, S. M. (1988). Cytoplasmic calcium measurements in intact higher plant cells: results from fluorescence ratio imagings of fura-2. *J. Cell Sci.* **91,** 71–80.
Cody, A. M., and Horner, H. T. (1985). Analytical resolution of the crystalline sand pyramids. *Am. J. Bot.* **72,** 1149–1158.
de Bary, A. (1887). "Comparative Morphology and Biology of the Fungi Mycetozoa and Bacteria." Clarendon Press, Oxford, England.
Dinsdale, D. D., Gordon, A. H., and George, S. (1979). Silica in the mesophyll cell walls of Italian rye grass (*Lolium multiflorum* Lam. cv. R.V.P.). *Ann. Bot.* **44,** 73–78.
Franceschi, V. R., and Horner, H. T. (1979). Use of *Psychotria punctata* callus in study of calcium oxalate crystal idioblast formation. *Z. Pflanzenphysiol.* **92,** 61–75.
Frank, E., and Jensen, W. A. (1970). On the formation of the pattern of crystal idioblasts in *Canavalia ensiformis* DC. IV. The fine structure of the crystal cells. *Planta* **95,** 202–217.
Garrity, D. P., Vidal, E. T., and O'Toole, J. C. (1984). Genotypic variation in the thickness of silica deposition on flowering rice spikelets. *Ann. Bot.* **54,** 413–421.
Goodwin, P. W., and Mercer, E. T. (1982). "Introduction to Plant Biochemistry," 2nd Ed., pp. 158–200. Pergamon, Oxford, England.
Hiltz, P., and Pobeguin, T. (1949). Sur la constitution des crystolithes de *Ficus elastica*. *C.R. Hebd. Seances Acad. Sci.* **228,** 1049–1051.
Hodson, M. J., and Sangster, A. G. (1987). Recent progress in botanical research on phytoliths. *Phytolitharien* **5,** 5–11.
Iler, R. K. (1978). Hydrogen-bonded complexes of silica with organic compounds. *In* "Biochemistry of Silicon and Related Problems" (G. Bendz and I. Linquist, eds.), pp. 53–76. Plenum, New York.
Jones, D., Wilson, M. J., and Tait, J. M. (1980). Weathering of a basalt by *Pertusaria corallina*. *Lichenologist* **12,** 277–289.
Jones, D., and Wilson, M. J. (1986). Biomineralization in crustose lichens. *In* Biomineralization in Lower Plants and Animals. B.S.C. Leadbeater and R. Riding, eds. Clarendon Press, Oxford, pp. 91–105.
Jones, D., Wilson, M. J., and McHardy, W. J. (1981). Lichen weathering of rock-forming minerals: Application of scanning electron microscopy and microprobe analysis. *J. Microsc. (Oxford)* **124,** 95–104.
Jones, L. H. P. and Handreck, K. (1967). Silica in soils, plants and animals. *Adv. Agron.* **9,** 107–149.
Jones, L. H. P., and Handreck, K. A. (1969). Uptake of silica by *Trifolium incarnatum* in relation to the concentration in the external solution and to transpiration. *Plant Soil* **30,** 71–80.
Jones, L. H. P., Milne, A. A., and Sanders, J. V. (1966). Tabashir: An opal of plant origin. *Science* **151,** 464–466.
Jones, R. L., and Hay, W. W. (1975). Bioliths. *In* "Soil Components: Inorganic Components" (J. E. Greseking, ed.), Vol. 2, pp. 481–496. Springer Verlag, New York.
Kaufman, P. B., Bigelow, W. C., Petering, L. B., and Drogasz, F. B. (1969). Silica in developing epidermal cells of *Avena* internodes: Electron microprobe analysis. *Science* **166,** 1015–1017.
Kaufman, P. B., Dayanandan, P., Takeoka, Y., Bigelow, W. C., Jones, J. D., and Iler, R. (1981). Silica in shoots of higher plants. *In* "Silicon and Siliceous Structures in Biological Systems" (T. L. Simpson and B. E. Volcani, eds.), pp. 409–449. Springer-Verlag, New York.

Kausch, A. P., and Horner, H. T. (1983). Development of syncytial raphide crystal ideoblasts in the cortex of adventitious roots of *Vanilla planifolia* L. (Orchidaceae). *Scanning Electron Microsc.* **2,** 893–903.

Ledbetter, M. C., and Porter, K. R. (1970). "Introduction to the Fine Structure of Plant Cells." Springer-Verlag, Berlin.

Lewin, J., and Reimann, B. E. F. (1969). Silicon and plant growth. *Annu. Rev. Plant Physiol.* **20,** 289–308.

Lin, W., Wagner, G. J., and Hind, B. (1977). The proton pump and membrane potential of intact vacuoles isolated from *Tulipa* petals. *Plant Physiol.* **59,** 85.

Lowenstam, H. A., and Weiner, S. (1983). Mineralization by organisms and the evolution of biomineralization. *In* "Biomineralization and Biological Metal Accumulation" (P. Westbroek and E. W. de Jong, eds.), pp. 191–203. Riedel, Dordrecht, The Netherlands.

McNair, J. B. (1932). The interrelation between substances in plants: Essential oils and resins, cyanogen and oxalate. *Am. J. Bot.* **19,** 255–272.

Macklon, A. E. S. (1984). Calcium fluxes at protoplasmalemma and tonoplast: Review. *Plant Cell Environ.* **7,** 407–413.

Macklon, A. E. S., and Sim, A. (1981). Cortical cell fluxes and transport to the stela in excised root segments of *Allium cepa* L. *Planta* **152,** 381–387.

Mandala, S., and Taiz, L. (1985). Proton transport in isolated vacuoles from corn coleoptiles. *Plant Physiol.* **78,** 104–109.

Mann, S., Perry, C. C., Williams, R. J. P. Fyfe, C. A., Gobbi, G. C., and Kennedy, G. J. (1983). The characterization of the nature of silica in biological systems. *J. Chem. Soc. Commun.* **1314,** 168–170.

Marty, F., Branton, D., and Leigh, R. A. (1980). Plant vacuoles. *In* "The Biochemistry of Plants" (N. E. Tolbert, ed.), Vol. 1, pp. 625–658. Academic Press, New York.

Matile, P. (1978). Biochemistry and function of vacuoles. *Annu. Rev. Plant Physiol.* **29,** 193–213.

Montgomery, D. J., and Wynn Parry, D. (1979). The ultrastructure and analytical microscopy of silicon deposition in the intercellular spaces of the roots of *Molinia caerulea* (L.) Moench. *Ann. Bot.* **44,** 79–84.

Okazaki, M., Ishida, Y., and Setoguchi, H. (1984). Calcification in higher plants and its physiological role with special reference to cystolith. *Bull. Tokyo Gakugei Univ. Ser. 4* **36,** 67–81.

Parry, D. W., Hodson, M. J., and Sangster, A. G. (1984). Some recent advances in studies of silicon in higher plants. *Philos. Trans. R. Soc. London, Ser. B* **304,** 537–549.

Perry, C. C., Mann, S., and Williams, R. J. P. (1984). Structural and analytical studies of the silicified macrohairs from the lemma of the grass *Phalaris canariensis* L. *Proc. Roy. Soc. Lond. B* **222,** 427–438.

Postek, M. T. (1981). The occurrence of silica in the leaves of *Magnolia grandiflora* L. *Bot. Gaz.* **142,** 124–134.

Ranson, S. L. (1965). The plant acids. *In* "Plant Biochemistry" (J. Bonner and J. E. Varner, eds.), pp. 443–525. Academic Press, New York.

Raven, J. A. (1983). The transport and function of silicon in plants. *Biol. Rev.* **58,** 179–202.

Sakai, W. S., and Sanford, W. G. (1984). A developmental study of silicification in the abaxial epidermal cells of sugarcane leaf blades using scanning electron microscopy and energy dispersive X-ray analysis. *Am. J. Bot.* **71,** 1315–1322.

Sakai, W. S., and Thom, M. (1979). Localization of silicon in specific cell wall layers of the stomatal apparatus of sugar cane by use of energy dispensive X ray analysis. *Anal. Bot.* **44,** 245–248.

Sangster, A. G. (1970a). Intracellular silica deposition in immature leaves in three species of the Gramineae. *Ann. Bot.* **34**, 245–257.

Sangster, A. G. (1970b). Intracellular silica deposition in mature and senescent leaves of *Sieglingia decumbens* (L.) Bernh. *Ann. Bot.* **34**, 557–570.

Sangster, A. G. and Hodson, M. J. (1986). Silica in higher plants. *In* "Silicon Biochemistry" (D. Evered and M. O'Connor, eds.), pp. 90–111. Wiley, Chichester (Ciba Foundation Symposium **121**).

Sangster, A. G., and Parry, D. W. (1971). Silica deposition in the grass leaf in relation to transpiration and the effect of dinitrophenol. *Ann. Bot.* **35**, 667–677.

Sangster, A. G., and Parry D. W. (1981). Ultrastructure of silica deposits in higher plants. *In* "Silicon and Siliceous Structures in Biological Systems" (T. L. Simpson and B. E. Volcani, eds.), pp. 383–407. Springer-Verlag, Berlin.

Saunders, J. A., and Conn, E. E. (1978). Presence of the cyanogenic glucoside Dhurrin in isolated vacuoles from *Sorghum*. *Plant Physiol.* **61**, 154–157.

Schumaker, K. S., and Sze, H. (1985). A Ca^{2+}/H^+ antiport system driven by the proton electrochemical gradient of a tonoplast H^+-ATPase from oat roots. *Plant Physiol.* **79**, 1111–1117.

Scurfield, G., Michell, A. J., and Silva, S. R. (1973). Crystals in woody stems. *Bot. J. Linn. Soc.* **66**, 277–289.

Scurfield, G., Anderson, C. A., and Segnit, E. R. (1974). Silica in woody stems. *Aust. J. Bot.* **22**, 211–230.

Simpson, T. L., and Volcani, B. E. (eds.) (1981). "Silicon and Siliceous Structures and Biological Systems." Springer-Verlag, New York.

Sterling, C. (1967). Crystalline silica in plants. *Am. J. Bot.* **54**, 840–844.

Sze, H. (1985). H^+-translocating ATPases: Advances using membrane vesicles. *Annu. Rev. Plant. Physiol.* **36**, 175–208.

Thom, M., and Komor, E. (1984). H^+–sugar antiport as the mechanism of sugar uptake by sugarcane vesicles. *FEBS Lett.* **173**, 1–4.

von Sachs, J. (1887). "Lectures on the Physiology of Plants." Clarendon Press, Oxford, England.

Wagner, G. J. (1977). Intracellular localization of vacuolar and cytosol components of protoplasts after vacuole isolation. *Plant Physiol. (Suppl.)* **59**, 104 (Abstr.).

Wagner, G. J., and Siegelman, H. W. (1975). Large-scale isolation of intact vacuoles and isolation of chloroplasts of mature plant tissues. *Science* **190**, 1298–1299.

Webb, M. A., and Arnott, H. J. (1982). A survey of calcium oxalate crystals and other mineral inclusions in seeds. *Scanning Electron Microsc.* **3**, 1109–1131.

Weiss, A., and Herzog, A. (1978). Isolation and characterization of a silicon–organic complex from plants. *In* "Biochemistry of Silicon and Related Problems" (G. Bendz and I. Lindquist, eds.), pp. 109–128. Plenum, New York.

Williams, R. J. P. (1978). Silicon in biological systems. *In* "Biochemistry of Silicon and Related Problems" (G. Bendz and I. Lindquist, eds.), pp. 561–576. Plenum, New York.

Williamson, R. E., and Ashley, C. C. (1982). Free Ca^{2+} and cytoplasmic streaming in the alga. *Chara. Nature (London)* **296**, 647–651.

Wilson, M. J., and Jones, D. (1984). The occurence and significance of manganese oxalate in *Pertusaria corallina* (Lichenes). *Pedobiologia* **26**, 373–379.

Wilson, M. J., Jones, D., and Russell, J. D. (1980). Glushinskite, a naturally occurring magnesium oxalate. *Mineral. Mag.* **43**, 837–840.

Zindler-Frank, E. (1976). Oxalate biosynthesis in relation to photosynthetic pathway and plant productivity—A survey. *Z. Pflanzenphysiol.* **80**, 1–13.

9 | Sponges—Spicules and Simple Skeletons

Of the multicellular animals that form skeletons, the marine sponges are the oldest and were present in the Proterozoic (1000 million years ago). The freshwater sponges, on the other hand, are much more recent, appearing in the Tertiary (50 million years ago) (Müller, 1982). By the middle Cambrian, three major classes had developed. However, it was not until the Carboniferous that the calcareous sponges appeared (Finks, 1970). Both marine and freshwater sponges form skeletons of calcium carbonate or of silica, but invertebrate groups that arose in later periods never again utilized silicon as the primary element in forming a mineral skeleton.

For the biologist, biochemist, and geologist, sponges claim attention on several counts. They assume an importance as a major component of the epifauna, as depositors of calcium carbonate and silica, and as eroders of rock and coral substrata. For the researcher with specific interests in skeletal formation, three more aspects can be added: (1) their cells have the capacity to deposit mineral by using two quite different methods, intracellular and extracellular; (2) the mineral units, the spicules, provide a stimulating challenge to understanding the cellular control of mineral deposition of a wide variety of shapes and dimensions and are the constructional units of elaborate arrangements; and (3) collagen, the organic portions of sponges, forms support in many taxonomic groups comparable to the interstitial tissue of vertebrates.

The biologists of the nineteenth century have provided excellent descriptions of sponge skeletons in all their varied forms and structure at the level of the light microscope. Interesting examples of studies of this period will be found in the Reports of the *H.M.S. Challenger* (Polejaeff, 1883), which provide beautifully illustrated descriptions of sponges collected on the voyages of this vessel during 1873–1876 (Fig. 9.1). Within the past three decades, the structural features of sponge cells and skele-

Figure 9.1 *1–2, Leucosolenia poterium* spermospore; *3,* L. *blanca* spicules and amoeboid cells; *4,* L. *challengeri,* trivadiate spicules; *5, Sycon arcticum,* horizontal section showing tubes, canals, and spicules; *6–13, Grantia tuberosa:* (6) horizontal section; (7) radial tube with spores; (8–13) spicules of various types.

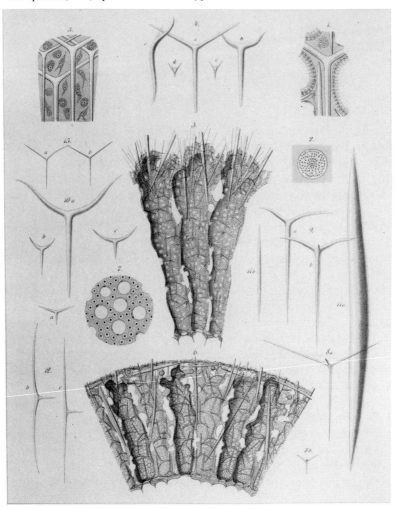

tons have been examined by transmission and scanning electron micros-copy, and experimental studies of spicule formation have been carried out. Useful sources of information are Grasse's *Traité de Zoologie, III. Spongiaires* (1973), Bergquist's *Sponges* (1978), and Simpson and Volcani's *Silicon and Siliceous Structures in Biological Systems* (1981).

The sponges have been divided into three main classes: the Calcarea, the Demospongiae, and the Hexactinellida. The taxonomic position of the fourth group, the Sclerospongiae, is still in discussion and it is in-cluded either with the Demospongiae or given class status. A brief sum-mary of the characteristics of the classes and their distribution is given in Table 9.1.

Table 9.1 Classes of Porifera[a]

Class	Habitat	Mineral Skeleton	Organic Skeleton
Calcarea	Marine	Calcium carbonate occurring as spic-ules or a fused mass. Spicules not differentiated as megascleres or microscleres	Collagen fibrils distributed hetero-geneously. Fibrils also surround spicules. Fibrils may be more abundant in areas of mechanical stress
Demospongiae	Fresh and brackish water. Marine from polar to tropical seas	Silica. Spicules 1- to 4-rayed. A few species have no spicules	Collagen may consti-tute part or entire skeleton or be absent. Collagen occurs as fibers or fibrils. A few species have no collagen
Hexactinellida	Marine, commonly deep sea	Silica. Megasclere and microsclere often fused and form a 3-dimen-sional network. Spicules have 3 axes which cross, forming 6 rays	No collagen. Tissue much reduced as compared with other classes
Sclerospongiae	Marine	Silica spicules and massive calcareous portion of arago-nite or calcite	Collagen fibers. Tissue reduced

[a] (After Bergquist (1978) and Garrone (1978).)

The basic structure of a sponge consists of three parts. An epithelium covers the exterior, and another epithelial layer lines the tubes through which water flows. Between the two epithelia is the mesohyl in which are various kinds of motile cells and the skeleton. The elements of the skeleton are a fibrous collagen and spicules of calcium carbonate or silica.

I. The Macroskeleton

The macroskeleton of most sponges is a supportive three-dimensional framework of spicules, of organic fibers without spicules, or a combination of fibers and spicules (Fig. 9.1). The exception is the coralline sponges, the Sclerospongiae, which have massive mineralized skeletons (Hartman and Goreau, 1970). That the organic material of the sponge skeleton falls within the collagen group of compounds was clearly demonstrated by Gross *et al.* (1956) by electron microscopy, chemical analysis, and X-ray diffraction. Their results have been amply confirmed and extended by later studies (Garrone, 1978). Three cell types secrete collagen: collencytes, lophocytes, and spongocytes (Bergquist, 1978). In

Figure 9.2 Portions of skeleton of *Haliclona oculata* (Pallas). (a) Siliceous spicules joined together with small blobs of spongin. (b) Siliceous spicules enclosed in thick spongin fibers (Hartman, 1981).

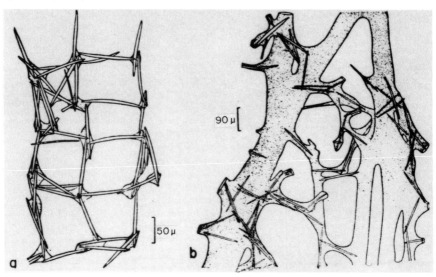

sponges with spicules, the association of collagen with spicules is an intimate one, and this collagen has been called spiculated spongin. It consists of fibers made up of fibrils and can be envisaged as a kind of matrix surrounding the spicules. The matrix : mineral ratio is greatest in that type of skeletal framework which takes the form of fibers with a small central core of spicules (Fig. 9.2b) and least in those instances in which the spongin is concentrated at junctions of the spicules (Fig. 9.2a). In another type of structure, the spongin serves to bind together bundles of spicules. It will be immediately apparent therefore that one of the functions of the spongin is to produce a coherent structural framework of some rigidity by binding together discrete mineral units (Hartman, 1981). However, in those freshwater sponges in which the joining is only at spicule junctions, the skeleton is a loosely knit structure (Garrone, 1978). A characteristic provided by spicule–spongin arrangements is flexibility of the framework. Lateral movement will be permitted in structures in which the spicules are associated in parallel yet separated by spongin. Here the fiber can move as a flexible rod. A skeleton constructed of a framework consisting of parallel fibrils which join rigid mineral units will resist and survive stresses from waves and moving sediment more effectively than a skeleton of fused spicules. Interestingly, the Hexactinellida, which have fused silica skeletons, live in deep water where wave and current stresses are reduced.

Spongin has a further function in attaching the sponge to its substratum. In those species in which the organic skeleton is well defined, the attaching spongin is continuous with the internal spongin (Garrone, 1978), and this is obviously advantageous in providing a holdfast which is not likely to separate from the organism when stress is applied. The firmness of anchoring to the substratum assumes particular importance in sponges living in areas of strong wave action such as the intertidal zone, shallow subtidal areas, and reefs.

II. The Mineral Skeleton

The mineral skeleton is formed from spicules that remain separate or are joined in arrangements that may be relatively complex. Individual spicules may be distributed within the body of the sponge, protrude from its surface (Fig. 9.1), or, as we have mentioned, be present within fibers of the organic skeleton. Spicules may also fuse to form a macroskeleton as in some Hexactinellida (Fig. 9.3) (Hartman, 1981), in genera of the Lithistida (Demospongiae), and in the Sclerospongiae. Be-

Figure 9.3   Hexactinellid skeletons. (a) Dictyonine framework of *Aphrocallistes bocagei* Wright. ×47. (b) Amphidisc of *Hyalonema sieboldii* Gray. ×834 (Hartman, 1981).

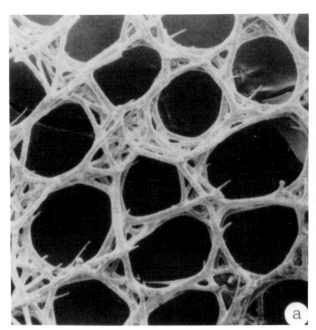

cause of spicule fusion, the skeleton of these sponges may be rigid and brittle. In orders and species lacking spicules, support is provided by the organic skeleton.

We mentioned that spicules are enormously varied in form. For convenience in describing them, various group and subgroup names are used denoting shape. Here are some examples: monaxons, rod-shaped; oxeas, monaxons with two similar pointed ends (Fig. 9.4b); triaxons or trienes (Fig. 9.4c), three axes as in the Hexactinellida; isochelas, two ends of varied form; asters (Fig. 9.4a), star-shaped with many points; and sterrasters, spherical with large numbers of units extending from a center (Fig. 9.5). Some spicules, although not all, can be divided into two large groups, the megascleres and the microscleres. As their names indicate, there is a marked difference in size of the two types. The length of megascleres ranges from 100 μm to several hundred micrometers. Microscleres are usually 10 to 100 μm in length (de Laubenfels, 1955). Both megascleres and microscleres are present in Demospongiae and

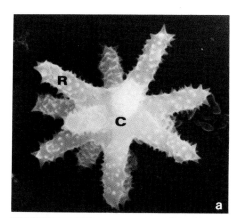

Figure 9.4 Scanning electron micrographs of silica spicules. (a) An aster showing the rays (R) which radiate out from the spicule center (C). All surfaces are heavily spined. ×5060). (b) An oxea, one type of larger spicule with pointed extremities. ×90. (c) A triene, the second type of larger spicule with one long, pointed ray and one branched extremity with three short rays (arrow). ×155 (Simpson *et al.*, 1985).

Figure 9.5 Sterraster of the tertractinomorph demo-
sponge *Geodia gibberosa* Lamarck. The depression marks
the position of the nucleus of the sclerocyte. ×2200 (Hartman, 1981).

Hexactinellida, but microscleres do not occur in Calcarea (Table 9.1). It is
the megascleres that are the main structural mineral units of sponge
skeletons, whereas the microscleres probably have no important support role beyond giving increased firmness to the tissues (Hartman,
1981).

III. Sponge Spicules and Their Formation

Sponge spicules are of two types, siliceous and calcareous. A considerable number of investigations have given attention to the fine structure of both types, and these studies have been summarized by Jones (1970, 1978) and Garrone *et al.* (1981).

A. Siliceous Spicules

Siliceous spicules consist of hydrated silica in a layered arrangement around an axis filament. Spicules with spines may have a filament within each spine as well (Simpson, 1981). Amino acid analysis of filaments indicates a protein rich in aspartic acid/asparagine but lacking the amino acids characteristic of collagen. Other compounds, still to be examined, may be present. The layering of silica spicules around the axis indicates either interruptions or changes during the course of deposition. Such layering, also referred to as incremental growth, is a widely occurring phenomenon in several invertebrate phyla. Among the many examples is the concentric layering of mineralized granules formed within intracellular vacuoles.

Both the microsclere and the megasclere spicules of siliceous sponges are formed intracellularly in special cells called sclerocytes (Fig. 9.6). These cells have a large nucleus, numerous mitochondria and microtubules, an active Golgi system, and many smooth-membraned vesicles. The spicule is formed within a special organelle bounded by a membrane. The first sign of spicule formation is the formation of an axial filament within this organelle. Its formation precedes mineralization, acts as a determinant of spicule shape, and serves as the material upon which silica deposition occurs (Garrone *et al.*, 1981). In some sclerocytes that form microscleres, there may be several vacuoles within a single cell. Each vacuole contains an individual spicule but the vacuoles may then fuse to form one large structure with up to 40 spicules all tightly packed together (Wilkinson and Garrone, 1980). When a siliceous spicule has been fully formed, a process that takes about 40 hr in the sponge *Ephydatia fluviatilis*, it punctures the sclerocyte, which then retracts. The spicule is then moved by a collencyte cell to its definitive position where additional cells may be involved in attaching it to other parts of the "skeleton," often with remarkable precision (Hartman, 1981).

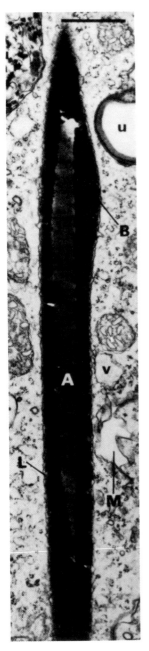

Figure 9.6 Longitudinal section of a spicule grown in pond water showing part of a bulb (B) which has been sectioned. Note the lower density of silica at the periphery of the bulb. M, Membranous pocket; u, inclusion of unknown function; v, vesicle; A, axial filament; L, silicalemma. ×22,300. Scale bar = 1.3 μm (Simpson and Vaccaro, 1974).

B. Calcareous Spicules

Calcareous spicules are impure calcite containing a solid solution of $MgCO_3$ (Jones, 1970, 1978). By placing spicules in a saturated solution of $Ca(HCO_3)_2$, they will be decorated with $CaCO_3$ crystals with an orientation relating to a crystal rhombohedron. Jones (1954) reported that, in slender monaxons of *Leucosolenia*, the optic axis coincides with the spicule axis and the crystal axis corresponds to the direction of growth. However, the relation of the optic axis cannot be applied generally to spicules since the optic angle does not correlate closely with the variations in spicule form seen in calcareous sponges (Jones, 1984). According to Jones (1978), there is no axial filament within calcareous spicules. Unlike siliceous spicules, the calcareous spicules of the Calcarea sponges are formed extracellularly through the action of two or more sclerocyte cells (Fig. 9.7). The sclerocytes completely envelop the initial spicule which is covered with a thin elastic sheath. The growing tips of the spicule are embedded in an invagination of the cytoplasm to form a dense cup. Around this region there are large numbers of vacuoles and invaginations in the cell membrane, apparently associated with the transport of calcium ions into this cavity. As the spicule grows, the sclerocytes become separated, leaving the spicule exposed (Jones, 1978). Simpson (1973) stated that secretion of calcareous spicules occurs intracellularly. The criterion was the presence of a membranous sheath on the spicule. Since the cells move over the spicule in depositing mineral, the presence of an organic coating is not surprising and does not in itself indicate an intracellular location. We favor the view that the calcareous spicules (but not siliceous spicules) are formed extracellularly. It appears that, in depositing the $CaCO_3$, the sclerocytes take in Ca^{2+} and probably HCO_3^- from the medium through their free surface and transport these ions through the cell membrane in contact with the spicule surface to the lattice surface. In the process, the sclerocytes must concentrate the Ca^{2+} and the HCO_3^- to the degree that $CaCO_3$ is formed. H^+ would necessarily be removed, presumably by passing into the cells. An exchange of H^+ and Ca^{2+} could perhaps take place across the sclerocyte membrane. It follows that the rate of $CaCO_3$ deposition on the spicule surface will be a function of ion transport.

As we have mentioned, sponges, in forming their spicules, have used two methods, extracellular for spicules of calcium carbonate and intracellular for spicules of silica. In both, the process can be divided into three phases. Nucleation and crystal growth begin the process. Then as growth continues, deposition must be controlled to give the characteris-

Figure 9.7 The development of calcareous sponge spicules. (a) Slender monact; (b) triact of *Clathrina coriacea* (Jones, 1970; all redrawn after Minchin, 1898, 1908).

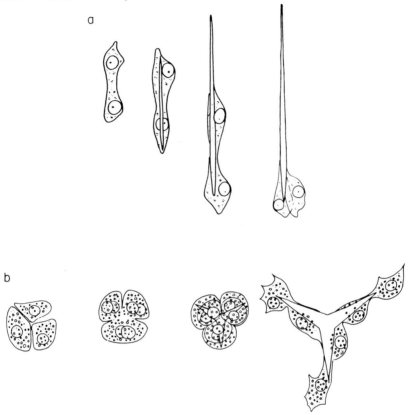

tic spicular form. Finally, growth is terminated as specific dimensions are attained in all planes.

IV. Unusual Aspects of Mineralization in Sponges

Spicule formation in sponges has not been studied in great detail and a number of observations would clearly warrant further investigation. Thus, Minchin (1910) commented that many of the older silica megascleres have "additional sclerocytes" associated with them along their length. The implication is that they may be involved in the later stages of silica deposition. The observation is interesting since silica deposition on the axial filament does not occur simultaneously along its entire length. Young spicules often have swellings during the early stages of development, suggesting either that these are sites of more efficient silicic acid transport or that there are specific sites of polymerization. The element germanium, which is similar chemically to silicon and can be incorporated into siliceous spicules, also enhances the swellings present in normal spicule development. In addition, germanium inhibits growth of the central filament and reduces the number of spicules formed. Its inhibitory action is irreversible (Simpson, 1981).

The discovery of the class Sclerospongiae (Hartman and Goreau, 1970) may stimulate interest in experimentation on the deposition of calcium carbonate and silica. These sponges have a basal skeleton of calcium carbonate as calcite or aragonite which is laid down around the siliceous spicules, indicating perhaps that there may be some common mechanisms in these rather different systems of mineralization. However, in some species, the siliceous and calcareous spicules are separate (Hartman, 1981).

V. Summary

1. The skeletons of sponges consist of mineral spicules and collagen fibrils. The collagen may enclose or unite the spicules into a flexible framework. In this respect, the fibrils are roughly comparable to the organic matrix of bryozoans, brachiopods, molluscs, and barnacles in which the organic matrix binds individual crystals into a macroscopic skeleton. Collagen fibers can also form a skeleton in the absence of spicules.
2. Sponges utilize two methods of mineralization: intracellular formation of siliceous spicules and extracellular formation of calcareous spicules. The deposition of silica involves movement of Si from the medium across the plasmalemma and vacuole membrane followed by polymerization within a vacuole. The formation of calcareous spicules appears to be accomplished through an active outward transport of calcium into the space between the plasmalemma and the spicule surface.
3. The sponges hold a special place in biomineralization in that they elaborate an enormous variety of spicule forms. The mechanism of control of form is largely unknown but genetic influence is evident in some taxa. Genetic influence is also expressed in the positioning of spicules by cells in the construction of the macroscopic skeletons.

4. Sponges have not acquired extracellular mineralization by epithelia, which has been important in the formation of exoskeletons of higher phyla. Instead, extracellular mineralization is carried out by single cells or groups of cells.

References

Bergquist, P. R. (1978). "Sponges." Univ. of California Press, Berkeley, California.

de Laubenfels, M. W. (1955). Porifera. *In* "Treatise on Invertebrate Paleontology," Part E, pp. E21–E112. Geological Society of America and the University of Kansas Press, Lawrence, Kansas.

Finks, R. M. (1970). The evolution and ecologic history of sponges during Palaeozoic times. *Symp. Zool. Soc. London* **25,** 3–22.

Garrone, R. (1978). Phylogenesis of connective tissue. Morphological aspects and biosynthesis of sponge intercellular matrix. *In* "Frontiers of Matrix Biology" (L. Robert, ed.), Vol. 5, pp. 1–250. Karger, Basel, Switzerland.

Garrone, R., Simpson, T. L., and Potter-Boumendel, J. (1981). Ultrastructure and deposition of silica in sponges. *In* "Silicon and Siliceous Structures in Biological Systems" (T. L. Simpson and B. E. Volcani, eds.), pp. 495–526. Springer-Verlag, New York.

Grassé, P. (1973). "Traité de Zoologie. III. Spongiaires." Masson et Cie, Paris.

Gross, J., Sokol, Z., and Rougie, M. (1956). Structural and chemical studies on the connective tissue of marine sponges. *J. Histochem. Cytochem.* **4,** 227–246.

Hartman, W. D. (1981). Form and distribution of silica in sponges. *In* "Silicon and Siliceous Structures in Biological Systems" (T. L. Simpson and B. E. Volcani, eds.), pp. 453–494. Springer-Verlag, New York.

Hartman, W. D., and Goreau, T. F. (1970). Jamaican coralline sponges: Their morphology, ecology and fossil relatives. *Symp. Zool. Soc. London* **25,** 205–243.

Jones, W. C. (1954). The orientation of the optic axis of spicules of *Leucosolenia complicata. Quart. J. Microsc. Soc.* **95,** 33–48.

Jones, W. C. (1970). The composition, development, form and orientation of calcareous sponge spicules. *Symp. Zool. Soc. London* **25,** 91–123.

Jones, W. C. (1978). The microstructure and genesis of sponge minerals. *Colloq. Nationaux C.N.R.S.* **291,** 425–445.

Jones, W. C. (1984). Spicule form in calcareous sponges (Porifera: Calcarea). The principle of uniplanar curvature. *J. Zool.* **204,** 571–584.

Minchin, E. A. (1898). Materials for a monograph of the ascons. I. On the origin and growth of the triradiate and quadradiate spicules in the family Clathrinidae. *Quart. J. Microsc. Sci.* **40,** 469–588.

Minchin, E. A. (1908). Materials for a monograph of the ascons. II. The formation of spicules in the genus *Leucosolenia,* with some notes on the histology of the sponges. *Quart. J. Microsc. Sci.* **52,** 301–335.

Minchin, E. A. (1910). Sponge-spicules. Summary of present knowledge. *Ergeb. Fortschr. Zool.* **2,** 171–174.

Müller, W. E. G. (1982). Cell membranes in sponges. *Int. Rev. Cytol.* **77,** 129–181.

Polejaeff, N. (1883). Calcarea. *In* "Reports of the Voyage of H.M.S. Challenger, 1873–76," Vol. 8. Neil, Edinburgh, Scotland.

Simpson, T. L. (1973). Coloniality among the Porifera. *In* "Animal Colonies" (R. S. Board-man, A. H. Cheetham, and W. A. Oliver, Jr., eds.), pp. 549–565. Dowden, Hutchin-son and Ross, Stroudsburg, Pennsylvania.

Simpson, T. L. (1981). Effects of germanium on silica deposition in sponges. *In* "Silicon and Siliceous Structures in Biological Systems" (T. L. Simpson and B. E. Volcani, eds.), pp. 527–550. Springer-Verlag, New York.

Simpson, T. L., and Vaccaro, C. A. (1974). An ultrastructural study of silica deposition in the freshwater sponge *Spongilla lacustris. J. Ultrastruct. Res.* **47,** 296–309.

Simpson, T. L., and Volcani, B. E. (eds.) (1981). "Silicon and Siliceous Structures in Biological Systems." Springer-Verlag, New York.

Simpson, T. L., Langenbruch, P.-F., and Scalera-Liaci, L. (1985). Silica spicules and axial filaments of the marine sponge *Stelleta grubii* (Porifera, Demospongiae). *Zoomorphology* **105,** 375–382.

Wilkinson, C. R., and Garrone, R. (1980). Ultrastructure of siliceous spicules and micro-sclerocytes in the marine sponge *Neofibularia irata. N. Sp. J. Morphol.* **166,** 51–64.

10

Echinoderms—Cells and Syncytia

Early in the Lower Cambrian period, the echinoderms devised a system for forming a mineral skeleton of macroscopic size that was both novel and new to the invertebrates. The bryozoans, brachiopods, and molluscs had all utilized an epithelial system of biomineralization which transfers ions, secretes proteinaceous material, and lays down compact mineral layers of small crystals. The echinoderm method departed from this and used instead a syncytium that deposited mineral. By this means, the cells fashioned skeletal units which were three-dimensional fenestrated structures of calcium carbonate rather than small calcium carbonate crystals. The deposition of mineral in a perforated pattern was not, however, solely an invention of the echinoderms. The single-celled radiolarians had mastered the art of making similarly elaborate mineral structures that are frequently as complex as some echinoderm units (frontispiece of book; Haeckel, 1887). Where the radiolarian and echinoderm skeletons differ is that the echinoderms construct multiunit structures by the collaboration of large numbers of cells.

The method of employing relatively large fenestrated mineral units made possible the formation of sizeable skeletons with three advantageous characteristics: they were light in weight, rigid, and resistant to fractures. Then, by making a body covering of many articulated mineral structures (Fig. 10.1) in which individual plates could grow at their borders and to which new plates could be added, the animal had the capacity for increasing in volume without the necessity of molting, which is such a hazardous activity in the Crustacea.

146

147

Figure 10.1 *Micropyga tuberculata.* Two views: *1,* seen from abactinal pole; *2,* seen from actinal side; *3,* actinal cut; *4,* interambulacral and ambulacral zones of actinal side; *5,* same from abactinal side; *6,* denuded abactinal system; *7,* interambulacral and ambulacral zones from abactinal side (*smaller specimen*); *8,* same from actinal side; *9,* abactinal system of same specimen magnified. (Agassiz, 1881).

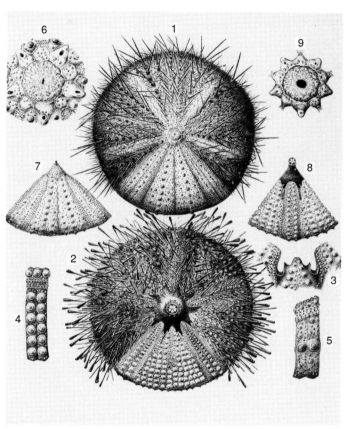

This skeletal meshwork or stereom (Fig. 10.5, 10.7) is present in the test, spines, and pedicellariae of sea urchins (class Echinoidea) and in the plates of brittle stars (class Ophiuroidea). In contrast, the sea cucumbers (class Holothuroidea) do not have a continuous mineral skeleton

Figure 10.2 (A) *Eucidaris tribuloides:* circumferential (compression) suture near the ambitus. (B) *E. tribuloides:* radial (tension) suture. Note dense stereom and equal amounts of fibers binding both sutures. (C–F) *Diadema antillarum:* (C) Junction between tension (transverse) and compression (vertical) sutures in interambulacral region; (D) compression joint between ambulacral plates, bound by collagen fibers; (E) compression joint between ambulacral plates broken open to show collagen fibers throughout thickness of joint; (F) tension suture between ambulacral plates broken open; note greater numbers of collagen fibers. All scale bars = 50 μm (Telford, 1985).

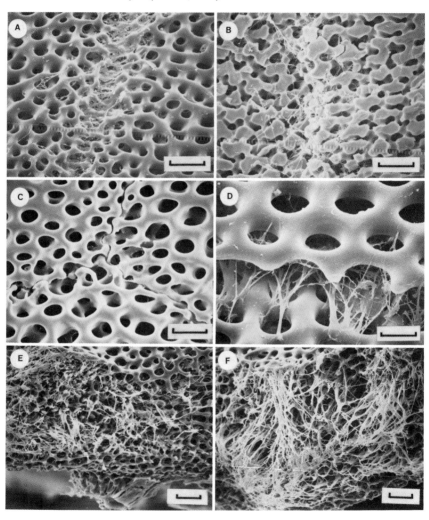

but possess an integument containing separate mineral ossicles which are often fenestrated and of microscopic size (Fig. 10.3).

This chapter will give attention primarily to the Echinoidea. Following an account of the mineral that is deposited, its physical properties, and the organic matrix within the mineral, we shall briefly describe the control of skeletal form and the method of mineralization.

Figure 10.3 (1–4) Different types of mineral ossicles of Holothuroidea. (5) Photomicrographs of living whole mounts of the body wall from several adult specimens of *Leptosynapta clarki* showing stages in the formation of a plate-like ossicle. (A) The double arrowheads mark a rod-shaped spicule deposited at the onset of ossicle formation near a well-developed anchor. (B–F) Subsequent stages in the branching and fusion of peripheral regions to form a fenestrated plate. Scale bar = 50 μm. ×300 (Stricker, 1986).

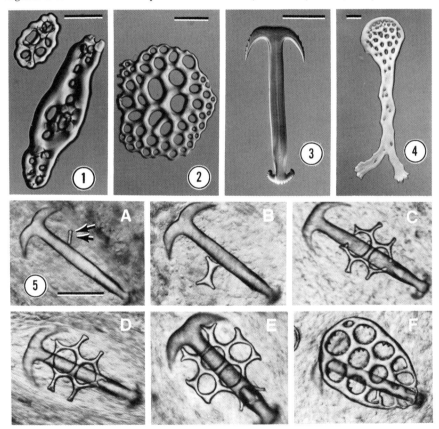

I. The Mineral System

The primitive form of the echinoid skeleton was spherical but from it an extensive array of body forms developed (Telford, 1985; Lawrence, 1987), including subspherical, hemispherical, dome-shaped, conical, globular, eggshaped, elongate, and flattened (Lawrence, 1987). These sea urchin skeletons are formed by mesenchyme cells and consist of four major structures (Pearse and Pearse, 1975). The main portion, called the test, is a mosaic of a large number of plates bound together by collagen fibers (Fig. 10.2). The spines are a second element, and these are anchored to the plates by muscles. The base of each spine and a portion of its plate form a ball-and-socket joint (Fig. 10.4), which makes it possible for them to move through very wide angles by contractions of the attached muscles. The tiny pedicellariae, which exist in various forms, are a third type of mineralized structure observed on the external surface of the test. The jaw apparatus in the central area of the ventral region of the test is the fourth major structure and consists of some 40 mineral units, including 5 teeth.

Figure 10.4 Scanning electron micrograph of the test of *Strongylocentrotus purpuratus*. Primary tubercle showing terminal knob (1), boss (2 and 3), and underlying plate (4). The spicule forms a ball-and-socket joint with the knob. ×140 (Okazaki *et al.*, 1981).

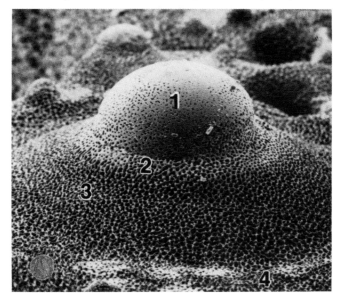

A. Test and Spines

Each plate of the test is seen in the scanning electron microscope to be a precisely constructed meshwork with smooth walls of calcite (Fig. 10.2). The channels or spaces of this stereom may account for more than 50% of the volume of the plates (Raup, 1966) and they provide an extensive microenvironment for the connective tissue cells and sclerocytes involved in mineral growth and repair throughout the skeleton. This system possesses advantageous physical properties, as we shall see later.

After examining the ultrastructure of 32 species of echinoids, Smith (1980) classified the plate stereoms into 10 basic types (Fig. 10.5). The type and form of the stereom may not be uniform, however, even within a single plate. For example, the inner part may be labyrinthic while the outer layer is a laminar stereom. Stereoms of a single type may also undergo changes as the plate becomes thicker (Pearse and Pearse, 1975; Smith, 1980). Apparently, the rate of growth may also influence the type of stereom that occurs within a plate. A moderate or fast growth rate may be associated with a labyrinthic stereom, whereas during slow growth a perforate stereom may be formed. The presence of tissue within the channels will also influence the microstructure of a stereom. Thus, connective tissue, collagen, and perhaps even the mineral-depositing sclerocytes may, by their presence, all have an influence on stereom structure.

Growth lines or bands are commonly present in the plates of many species and their occurrence is invariably associated with differences in the structure of the stereom (Pearse and Pearse, 1975; Smith, 1980). This, in turn, has been correlated with seasonal changes in growth rate and with such physiological influences as gonad development and spawning.

A stereom structure is also found in echinoid spines but it has quite a different pattern from the test plates. A transverse section of a spine of *Strongylocentrotus*, for example, shows a concentric arrangement of rings called cycles, which in longitudinal section appear as bands, each cycle representing a growth increment (Fig. 10.6a) (Heatfield, 1971). The spine structure varies with the species, as can clearly be seen by comparing the spine of *Strongylocentrotus* with a spine of *Diadema*, which is a hollow structure with a perforated calcite cylinder (Fig. 10.7) (Burkhardt *et al.*, 1983).

B. Teeth

Teeth are made up of two rows of twin-bladed calcareous units stacked closely together (Fig. 10.8). Each of the two main parts of the tooth has

Figure 10.5 Complete block diagrams of stereom fabrics. From Smith (1980).

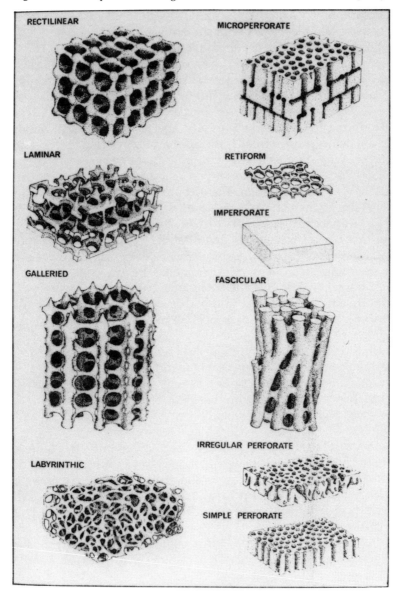

Figure 10.6 Photomicrographs of the internal structure of the skeleton of primary interambulacral spines of *Strongylocentrotus*. Portions of a transverse section of the shaft about 3 mm above the milled ring: (a) From the inner zone of meshwork (M) to the edge of the outer zone of wedges. A and B are 2 of the 12 wedges shown. Four cycles are present as indicated by arrows. (b) Higher magnification showing the outer cycle (between arrows) of two wedges, A and B. (Br, Bridge; CA, canals; M, meshwork between adjacent wedges. Scale line = 0.2 mm (Heatfield, 1971).

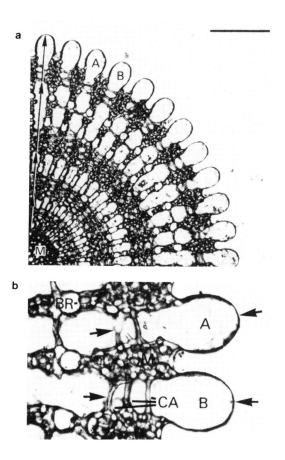

Figure 10.7   Skeleton of a primary spine of *Diadema setosum*: longitudinal section of the base. (Burkhardt *et al.*, 1983).

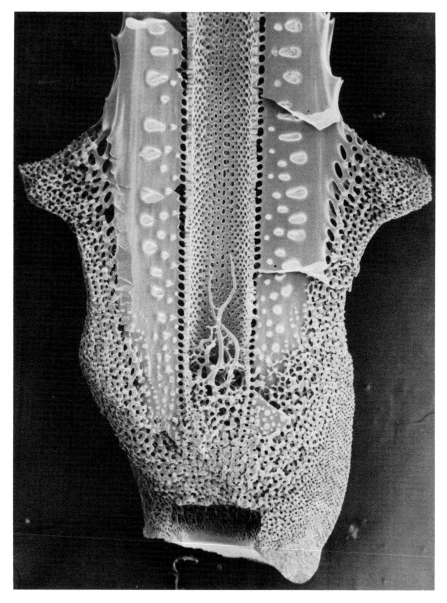

Figure 10.8   Teeth and lanterns of a regular sea urchin with keeled teeth (a) and of a sand dollar (b). a_1, b_1: longitudinal section through test and lantern; a_2, b_2: horizontal sections through the lanterns; a_3, b_3: side view of the teeth. (Note the differing size of the plumulae.) a_4, b_4: transversal cuts through the teeth. HV numbers are Vickers hardness numbers (Märkel *et al.*, 1977; after Märkel and Gorny, 1973, with changes). No scale.

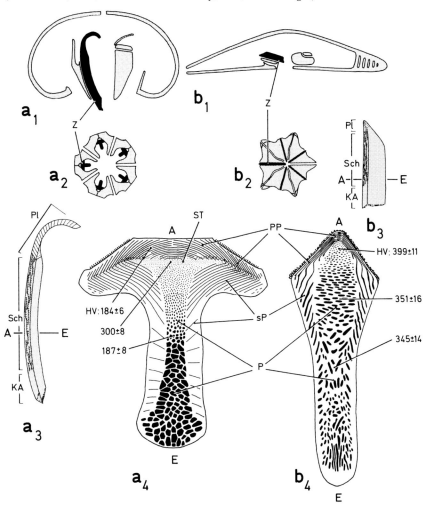

primary plates (PP), side plates (SP), and elongate fibers of calcium carbonate called prisms (P). These various structures are formed continuously by odontoblast cells at the base of the tooth (Greer and Weber, 1969; Märkel *et al.*, 1977, 1986). Calcareous disks present in the shaft of the tooth cement the plates into a unified structure. The primary plates and the side plates are monocrystalline. The calcareous disks are poly-crystalline. One limited area of the tooth, the stone zone, has extreme hardness and consists of short calcareous fibers. The magnesium content of this region has been found to be 38–43.5 mmol/100 g (Schroeder *et al.*, 1969; Märkel *et al.*, 1977) and is accounted for by the magnesium content of the calcareous disks.

The teeth of sand dollars are structurally different from the teeth of sea urchins, and these differences are correlated with functional differences between the two types of teeth (Märkel *et al.*, 1977). The five sand dollar teeth are arranged almost horizontally in a circle and exert a crushing force through muscles which move the teeth horizontally toward the center of the circle. The crushing pressure is accordingly along the long axis of the tooth. The teeth of sea urchins, in contrast, protrude from the mouth and are chisel-like scrapers rather than crushers. Their edges also move inward toward the center of the ring of teeth, but as the sharp edges of the tooth scrape over the surface, the tooth is subjected to unevenly distributed stress (Märkel *et al.*, 1977).

II. Echinoderm Mineral

A. Physical Properties

Echinoderm mineral is magnesian calcite but the crystalline nature of echinoderm skeletal units has long been controversial. The question has been whether the mineral units are a single crystal or whether they are made up of crystallites. Under polarized light the spines, primary elements of teeth, and some, but not all, plates appear to be single crystals (Raup, 1966; Okazaki *et al.*, 1981). Evidence that spines are single crystals also comes from X-ray diffraction (Donnay and Pawson, 1969). By etching and crystal decoration techniques, the spines of *Strongylocentrotus purpuratus* were found to have common *a*- and *c*-axes in all cross sections (Okazaki *et al.*, 1981). This uniformity in two crystal axes extends earlier findings of a common *c*-axis and provides increased support for the view that the spine could be considered a single crystal. The ambulacral plates, on the other hand, were found to have different *a*- and *c*-axes in various parts of a single plate. The tubercles, which are the bases on which the spines rotate, appeared as polycrystalline aggregates. This information on the diversity of crystal forms was extended by the X-ray data of Garrido and Blanco (1947) and Nissen (1963), who suggested that the skeletal units may be composed of tiny crystallites. A somewhat similar conclusion was reached by Towe (1967) who also considered that echinoderm mineral may be a "mosaic" of crystallites. If such crystallites were in perfect or nearly perfect alignment, the skeletal unit would, of course, appear to be a single crystal when observed by polarized light or X rays. To that extent, the definition of a single crystal depends on the resolving power of the instruments being used.

Support for the presence of crystallites within skeletal units has come from selected area electron diffraction (Blake and Peacor, 1981) and the appearance of natural (Towe, 1967) and fracture surfaces (O'Neill, 1981). In the latter case, ossicles of the sea star *Echinaster spinulosus,* when fractured after loading in stress relaxation, showed needlelike crystallites in a parallel orientation. Recently, lattice fringe images from high-resolution transmission electron microscopy have provided evidence of "mosaic blocks" or crystallites in skeletal units of the crinoid *Neocrinus blakei* (Blake *et al.,* 1984). It is obvious therefore that this structure of magnesian calcite differs from nonbiological magnesian calcite. Clearly, echinoderm skeletal units do not represent physicochemical precipitation, but show the influence of the mineralizing cells which form them. This is not particularly surprising since it is likely that these cells release organic material which could well influence the mineral ultrastructure. How that interaction occurs, however, is an important problem in molecular biology.

B. Resistance to Stresses

The resistance of the echinoid test to stresses will depend on its form, the structure of its plates, and the way that they are bound together at the sutures. As we have mentioned, one form of test is dome-shaped. The forces acting on such tests are, as a first approximation, similar to those in domes of buildings. They are of two kinds, compressive and tensile (Telford, 1985). The compressive forces act radially and tend to flatten the dome (Fig. 10.9a). The tensile forces will be circumferential and have the effect of increasing the dome's diameter (Fig. 10.9b). In contrast to the dome of a building, the weight of the test with its low specific gravity will not be of especial importance in this context, whereas loading forces encountered in the environment may be of considerable significance.

Test plate structure and interplate binding bring other factors into play in accommodating the compressive and tensile forces (Telford, 1985). The plates are usually bound together by collagen fibers (Fig. 10.2), and this permits slight movements in response to stresses. In some species, however, the sutures are reinforced by fusion of the trabeculae of contiguous plates. Another factor is the difference in test

ratio greater than 1.0 in the skeletal $CaCO_3$ indicates that metabolic CO_2 is being converted to carbonate and so is reducing the contribution of dissolved inorganic ^{14}C from the medium. Since cells have an ionic calcium content of about $10^{-7} M$, we can safely assume that the cells make no contribution that would appreciably alter the $^{45}Ca/^{14}C$ ratio of deposited $CaCO_3$. This method has been applied in measurements of rates of simultaneous uptake of ^{45}Ca and dissolved inorganic ^{14}C in the prism stage of larvae of *Strongylocentrotus purpuratus* (Sikes *et al.*, 1981). The $^{45}Ca/^{14}C$ ratio of the calcium carbonate deposited in the skeleton was 2.7. Thus, for each ^{14}C ion coming from the medium, 1.7 ions of carbonate were supplied by metabolic $^{12}CO_2$.

The echinoids have been found to have carbonic anhydrase in larvae and in many parts of the adults (see Chen and Lawrence, 1986; J. E. Donachy, unpublished results). The enzyme may well facilitate the uptake of bicarbonate from the medium and the movement of carbon dioxide and bicarbonate into the calcifying tissues. There is a decreased uptake of $NaH^{14}CO_3$ into teeth of *Lytechinus in vitro* in the presence of the enzyme inhibitor acetazolamide. This suggests that the carbonic anhydrase may be important in facilitating the movement of the anion through the tissues (Chen, 1985).

V. Cells and Cellular Cooperation

The complexity of the echinoderm skeleton, composed as it is of plates, ossicles, teeth, and spines, immediately raises the question of how the deposition of such a detailed structure can be controlled. The question is extremely difficult to answer since it poses both technical and intellectual difficulties. Fortunately, the echinoderm system of biomineralization is sufficiently unusual to provide insights that would be hard to envisage in other phyla. Biomineralization in the echinoderms involves both motile cells and syncytia (Märkel *et al.*, 1986), and the simplest explanation of what is happening can be obtained from studies of the regeneration of the test and spines of sea urchins.

When the spine of an urchin such as *Strongylocentrotus purpuratus* is broken, the base regenerates and provides an excellent system for studying a whole range of physiological influences (Heatfield, 1971). At the cellular level, the process involves mobile sclerocyte cells. The mineralization process starts with the formation of an oriented crystal within a cytoplasmic vacuole. These intracellular crystals seem to grow in association with the Golgi complex and mitochondria. As the vacuole increases in size, it eventually ruptures and the surface of the crystal is then covered by other cells, the secondary sclerocytes, which continue to secrete the material of the regenerating spine (Shimizu and Yamada, 1980). The process is illustrated in Fig. 10.10, which clearly demonstrates intracellular mineralization and the subsequent cooperation between secondary sclerocytes. Here then is a system involving both intracellular and extracellular mineralization.

In the sea urchin gastrula, cell division forms a mesenchymal aggregate of cells. These cells produce pseudopodial extensions which fuse and form a syncytial center. The pseudopodial extensions pass from this region to the original cells, forming what is referred to as the "pseudopodial cable." The larval skeleton is formed within a vacuole of this syncytial mass. According to Okazaki and Inoue (1976), the mesenchyme cells may change the shape of the vacuole and in this way determine the axes of the developing spicule. The spicule becomes triradiate, the arms oriented in directions corresponding to the axes of a calcite crystal (Fig. 10.11). In fact, the form of the spicule is such that it could have been carved from a single crystal of calcite. By immersing the

Figure 10.10 Schematic illustration of the general course of crystal formation in regenerating test and spine of *Strongylocentrotus intermedius*. (c) Crystal (Shimizu and Yamada, 1980).

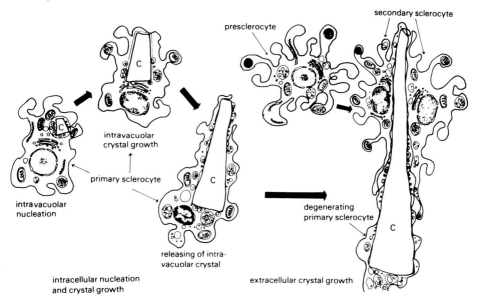

Figure 10.11 Pluteus of *Arbacia punctulata* fixed by methanol, mounted in Canada balsam, and observed with a polarizing microscope. A pair of skeletal spicules shows high negative birefringence. Despite its complex morphology, the whole left (C) or right (A) spicule is extinguished when the body rod becomes oriented parallel to the polarizer or analyzer axis. The optic axis of the spicule lies along the body rod. Optically, the left and right spicule behave as though carved out of a single crystal of calcite (Okazaki and Inoue, 1976).

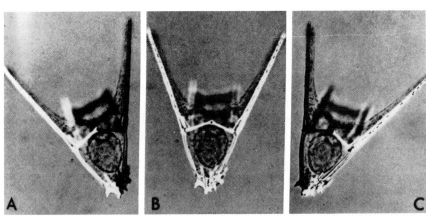

spicules in 0.1 M CaCl$_2$ and 0.1 M NaHCO$_3$, they become decorated with microcrystals, each identical to a cleavage rhombohedron of calcite and all parallel to each other. This method demonstrates that the triradiate arms are parallel to the three crystal axes of calcite, as their angles had also indicated.

The growth of the larval skeleton beyond the simple triradiate spicules is accomplished by the addition of further mesenchyme cells derived from micromeres. Any cell in this syncytial arrangement may leave and reenter by severing and reforming the pseudopodial links. Cells are therefore able to change their location and contribute to further growth throughout the development of the spicule. When the cells in the syncytium divide, they temporarily sever their connections with the rest of the structure (Okazaki *et al.*, 1980).

The spicules that are formed in this way have shapes that are characteristic of each species and, as such, differ from normal crystalline calcite. Clearly, these features must be determined biologically.

If a population of micromeres is isolated from the dividing sea urchin egg at the 16-cell stage, they can be maintained in culture and will eventually form skeletal rods (Okazaki, 1975). In the process of mineral deposition, single mineralizing cells can be seen scattered along the rods

(Fig. 10.12). Since the parts of the rods have a species-specific conformation, the individual cells must possess specific information as to the placing of bends and branches. The fact that spicule shape is genetically controlled can also be shown by crossing the two sea urchins *Lytechinus variegatus* and *Tripneustes esculentis*. The hybrid pluteus larva produces abnormal spicules showing some features of both parents (McClay and Hausman, 1975).

Once the species-specific pattern is established, it is clearly maintained by the sclerocytes. These cells are apparently programmed to cease deposition when they have reached the appropriate size, just as they are programmed to fashion appropriate curvatures. The method by which specific genes bring about the cellular control of deposited mineral by populations of cells is one of the central challenges of biology; and the echinoderm system of biomineralization is, as we have already seen, ideal material for its study.

Similar opportunities for experimentation exist in the development of the fenestrated ossicles present in the outer covering of the holothurian sea cucumbers (Fig. 10.3A–F) (Stricker, 1986). A rodlike spicule first forms. It then grows arms which elongate, branch, and fuse at their ends, finally producing a platelike structure with holes. In the Holothuroidea, as in the Echinoidea, the pattern of fenestrations is species-specific and the cells which form the plates and the ossicles are sclerocytes which are united in syncytia.

A related phenomenon is found in the formation of the teeth of the echinoid *Lytechinus variegatus*. The top of the tooth is enclosed by a single layer of cells. Individual odontoblasts migrate from there to one surface where they again fuse to form a syncytium. The solitary odontoblasts have patchy chromatin, little cytoplasm, and show very limited activity until they fuse. They then develop an abundance of endoplasmic reticulum, numerous vesicles, and a conspicuous Golgi system so that the syncytium shows all the requirements for the mineralizing activities that it then assumes (Chen and Lawrence, 1986). The subsequent growth of the tooth is especially interesting in that many mineral units of two types are deposited individually and are then cemented together to form the completed tooth structure. Details of this construction are not fully known, but the initial events have been set forth in an hypothesis by Märkel *et al.* (1986) based on observations of *Eucidaris tribuloides*. Each of the mineral units of a tooth is formed in a vacuole within the syncytial sheath of the odontoblasts (Fig. 10.13). The unit is coated with organic material apparently originating within cells of the syncytium and indicating an intimate relation between the organic material and the

Figure 10.12

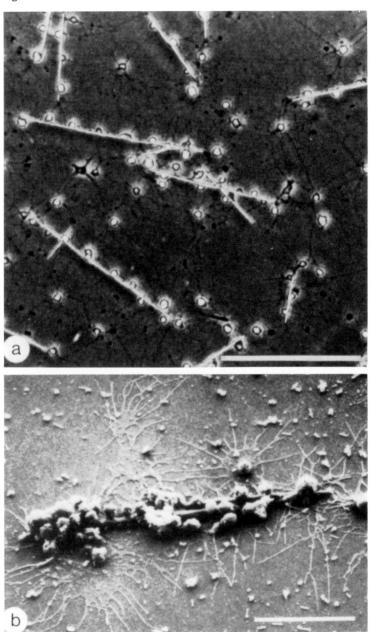

Figure 10.12 (*facing page*)   Primary mesenchyme cells cultured *in vitro*. (a) Light micrograph (phase mode) of spicules and associated cells cultured *in vitro* for 72 hr in seawater containing horse serum. (b) Scanning electron micrograph revealing spicule enclosed in the filopodia-derived sheath. Bars: (a) 205 μm, (b) 55 μm (Decker and Lennarz, 1988).

Figure 10.13   Schematic diagram summarizing the proposed hypothesis on calcite deposition of mineral units of a tooth in echinoderms. The calcite plate is enlarged on the left of the diagram and full-grown to the right. Not to scale (Märkel *et al.*, 1986).

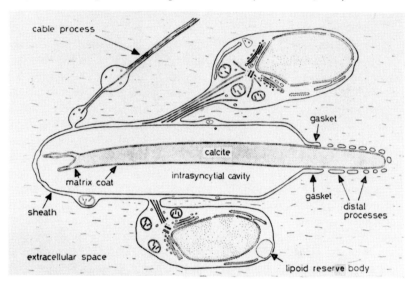

mineral. The form of the mineral unit is presumed to be dictated by the syncytium which serves as a mold. During the assembly of the teeth, the odontoblasts carry out a second step in which calcitic disks are deposited and then joined with the first-formed units to provide structural unity to the tooth. Mineral deposition by syncytia, long known to be the method of forming larval spicules, appears to be the mechanism of forming echinoderm teeth and ossicles as well.

VI. Summary

1. The echinoderms initiated a system of mineralization that made possible the construction of skeletons consisting of relatively large mineral units. This involves the mineralizing activity of syncytia rather

than the epithelial deposition of small crystals in layered arrangements that is present in several other invertebrate phyla.

2. An innovation of echinoderm evolution associated with the large size
   of the mineral units was their fenestrated construction. Each unit or
   plate of the test could then become a self-contained growth chamber
   of mineralizing cells, with the result that each unit could grow independently of other units. The echinoderm test formed of a mosaic of
   many plates could then become larger through the mineral deposition of its component plates.

3. Echinoderm cells have three methods of mineralization. (a) The mineral deposition in vacuoles within syncytia forms the larval skeleton,
   ossicles, and tooth units in addition to the early formation of the
   plates of the test. (b) Intracellular mineralization is used to initiate the
   skeletal repair of test and spines. (c) Extracellular mineralization by
   individual cells is probably responsible for the normal growth of
   plates of the test and the repair of both the test and spines.

4. The ability of syncytia to fashion the species-specific form of larval
   skeletons demonstrates the precise genetic influence that occurs in
   the mineralization processes. This influence is also evident in the
   shape of the fenestrations in units of the test and in the form of
   mineral rods of larval skeletons deposited by individual cells in
   culture.

5. The mineral of the test and teeth has an organic matrix within it that
   consists of a soluble and insoluble fraction. The matrix of spicules of
   the larval skeleton includes several proteins of 67 kDa or less. The
   functions of the components of this skeletal matrix are currently being explored and provide an exciting insight into the molecular biology of the relationships of the organic compounds of the matrix to
   mineral deposition and to the final form of the skeletal units.

References

Agassiz, A. (1881). Echinoidea. In "Report of the Scientific Results of the Voyage of H.M.S.
   Challenger, 1873–76," Vol. III. Neil, Edinburgh, Scotland.

Benson, S. C., Benson, N. C., and Wilt, F. (1986). The organic matrix of the skeletal spicule
   of sea urchin embryos. J. Cell Biol. 102, 1878–1886.

Benson, S., Sucov, H., Stephens, L., Davidson, E., and Wilt, F. (1987). A lineage-specific
   gene encoding a major matrix protein of the sea urchin embryo spicule. I. Authentica-
   tion of the cloned gene and its developmental expression. Dev. Biol. 120, 499–506.

Berman, A., Addadi, L., and Weiner, S. (1988). Interactions of sea urchin skeleton macromolecules with growing calcite crystals—A study of intracrystalline proteins. Nature
   (London) 331, 546–548.

Blake, D. F., and Peacor, D. R. (1981). Biomineralization in crinoid echinoderms: Characterization of crinoid skeletal elements using TEM and STEM microanalysis. *Scanning Electron Microsc.* **3**, 321–328.

Blake, D. F., Peacor, D. R., and Allard, L. F. (1984). Ultrastructural and microanalytical results from echinoderm calcite: Implications for biomineralization and diagenesis of skeletal material. *Micron. Microsc. Acta* **15**, 85–90.

Burkhardt, A., Hansmann, W., Märkel, K., and Niemann, H. J. (1983). Mechanical design of spines of diadematoid echinoids (Echinodermata, Echinoidea). *Zoomorphology* **102**, 189–202.

Chen, C. P. (1985). "The Role of Carbonic Anhydrase in the Calcification of the Tooth of the Sea Urchin *Lytechinus variegatus* (Lamarck)," Ph.D. thesis. Univ. of South Florida, Tampa, Florida.

Chen, C. P., and Lawrence, J. M. (1986). Localization of carbonic anhydrase in the plumula of the tooth of *Lytechinus variegatus* (Echinodermata: Echinoidea). *Acta Zool.* **67**, 27–32.

Decker, G. L., and Lennarz, W. J. (1988). Skeletogenesis in the sea urchin embryo. *Development* **103**, 231–247. Company of Biologists, Ltd.

Donnay, G., and Pawson, D. L. (1969). X-Ray diffraction studies of echinoderm plates. *Science* **166**, 1147–1150.

Emlet, R. B. (1982). Echinoderm calcite: A mechanical analysis from larval spicules. *Biol. Bull. (Woods Hole, Mass.)* **163**, 264–275.

Garrido, J., and Blanco, J. (1947). Structure cristalline des piquants d'oursin. *C.R. Hebd. Seances Acad. Sci.* **224**, 485.

Greer, R. T., and Weber, J. N. (1969). Comparative morphology of the echinoid tooth. *Scanning Electron Microsc.* 1969, 157–168.

Haeckel, E. (1887). Report on the Radiolaria collected by *H.M.S. Challenger* during the years 1873–1876. *In* "Reports of the Voyage of *H.M.S. Challenger*," Vol. 18. Neil, Edinburgh, Scotland.

Heatfield, B. M. (1971). Growth of the calcareous skeleton during regeneration of spines of the sea urchin, *Strongylocentrotus purpuratus* (Stimpson): A light and scanning electron microscopic study. *J. Morphol.* **134**, 57–90.

Kitajima, T. (1986). Differentiation of sea urchin micromeres: Correlation between specific protein synthesis and spicule formation. *Dev. Growth Differ.* **28**, 233–242.

Klein, L., and Currey, J. D. (1970). Echinoid skeleton: Absence of a collagenous matrix. *Science* **169**, 1209–1210.

Lawrence, J. M. (1987). "A Functional Biology of Echinoderms." The Johns Hopkins Univ. Press, Baltimore, Maryland.

McClay, D. R., and Hausman, R. E. (1975). Specificity of cell adhesion: Differences between normal and hybrid sea urchin cells. *Dev. Biol.* **47**, 454–460.

Märkel, K., Gorny, P., and Abraham, K. (1977). Microarchitecture of sea urchin teeth. *Fortschr. Zool.* **24**, 103–114.

Märkel, K., and Gorny, P. (1973). Zur funktionellen Anatomie der Seeigelzähne. *Z. Morph. Tiere* **75**, 223–242.

Märkel, K., Roser, U., Mackenstedt, U., and Klostermann, M. (1986). Ultrastructural investigation of matrix-mediated biomineralization in echinoids (Echinodermata, Echinoidea). *Zoomorph.* **106**, 232–243.

Nissen, H. U. (1963). Crystal orientation and plate structure in echinoid skeletal units. *Science* **166**, 1150–1152.

Okazaki, K. (1975). Spicule formation by isolated micromeres of the sea urchin embryo. *Am. Zool.* **15**, 567–581.

Okazaki, K., and Inoue, S. (1976). Crystal property of the larval sea urchin spicule. *Dev. Growth Differ.* **18**, 413–434.

Okazaki, K., McDonald, K., and Inoue, S. (1980). Sea urchin larval spicule observed with the scanning electron microscope. *In* "The Mechanisms of Biomineralization in Animals and Plants" (M. Omori and N. Watabe, eds.), pp. 159–168. Tokai Univ. Press, Tokyo, Japan.

Okazaki, K., Dillaman, R. M., and Wilbur, K. M. (1981). Crystalline axes of the spine and test of the sea urchin *Strongylocentrotus purpuratus:* Determination by crystal etching and decoration. *Biol. Bull. (Woods Hole, Mass.)* **161**, 402–415.

O'Neill, P. L. (1981). Polycrystalline echinoderm calcite and its fracture mechanics. *Science* **213**, 646–648.

Pearse, J. S., and Pearse, V. B. (1975). Growth zones in the echinoid skeleton. *Am. Zool.* **15**, 731–753.

Pilkington, J. B. (1969). The organization of skeletal tissues in the spines of *Echinus esculentus. J. Mar. Biol. Assoc. U.K.* **49**, 857–877.

Raup, D. M. (1966). The endoskeleton. *In* "Physiology of Echinedormata" (R. A. Boolootian, ed.), pp. 379–395. Wiley (Interscience), New York.

Schroeder, J. H., Dwornik, E. J., and Pipiker, J. J. (1969). Primary protodolomite in echinoid skeleton. *Geol. Soc. Am. Bull.* **80**, 1613–1616.

Shimizu, M., and Yamada, J. (1980). Sclerocytes and crystal growth in the regeneration of sea urchin test and spines. *In* "The Mechanisms of Biomineralization in Animals and Plants" (M. Omori and N. Watabe, eds.), pp. 169–178. Tokai Univ. Press, Tokyo, Japan.

Sikes, C. S., Okazaki, K., and Fink, R. D. (1981). Respiratory CO_2 and the supply of inorganic carbon for calcification of sea urchin embryos. *Comp. Biochem. Physiol. A* **70**, 285–291.

Smith, A. B. (1980). Stereom microstructure of the echinoid test. *Spec. Pap. Palaeontol.* **25**.

Stricker, S. (1986). The fine structure and development of calcified skeletal elements in the body wall of holothurian echinoderms. *J. Morphol.* **188**, 273–288.

Sucov, H. M., Benson, S., Robinson, J. J., Britten, R. J., Wilt, F., and Davidson, E. H. (1987). A lineage specific gene encoding a major matrix protein of the sea urchin embryo spicule. II. Structure of the gene and derived sequence of the protein. *Dev. Biol.* **120**, 507–519.

Swift, D. M., Sikes, C. S., and Wheeler, A. P. (1986). Analysis and function of organic matrix from sea urchin tests. *J. Exp. Zool.* **240**, 65–74.

Telford, M. (1985). Domes, arches and urchins: The skeletal architecture of echinoids (Echinodermata). *Zoomorphology* **105**, 114–124.

Towe, K. M. (1967). Echinoderm calcite: Single crystal or polycrystalline aggregate. *Science* **157**, 1048–1050.

Veis, D. J., Albinger, T. M., Clohisy, J., Rahima, M., Sabsay, B., and Veis, A. (1986). Matrix proteins of the teeth of the sea urchin *Lytechinus variegatus. J. Exp. Zool.* **240**, 35–46.

Venkatesan, M., and Simpson, R. T. (1986). Isolation and characterization of spicule proteins from *Strongylocentrotus purpuratus. Exp. Cell Res.* **166**, 259–264.

Wainwright, S. A., Biggs, W. D., Currey, J. D., and Gosline, J. M., eds. (1976). "Mechanical Design in Organisms." Arnold, London.

Weiner, S. (1983). Mollusk shell formation: Isolation of two organic matrix proteins associated with calcite deposition in the bivalve *Mytilus californianus. Biochemistry* **22**, 4139–4144.

Weiner, S. (1985). Organic matrixlike macromolecules associated with the mineral phase of sea urchin skeletal plates and teeth. *J. Exp. Zool.* **243**, 7–15.

11

Coelenterates—Epithelia, Symbiotic Influences, and Energy Metabolism

The coelenterates, or more precisely the members of the phylum Cnidaria, are a totally aquatic group of animals. Only a few have adapted to freshwater and all the mineralizing forms are marine. These include the reef-building stony corals, the soft corals, sea whips, sea fans, and sea pens. In discussing them we will concentrate mainly on the sea whips (Gorgonacea) and the stony corals (Scleractinia) (Table 11.1).

The coelenterates have a characteristic level of organization consisting of two epithelia, representing an outer ectoderm and an inner endoderm. At its simplest, the animal is a two-layered sac in which the two epithelia are separated by a glycoprotein secretion, the mesoglea, which forms a gel that can be distorted (Fig. 11.1). Given this basic anatomical plan and the often observed association between biomineralization and a glycoprotein matrix, one might expect to find the mesoglea being calcified to form an internal skeleton. In fact, this never happens. The mesoglea always remains as a flexible supporting system and, when it is strengthened, it is by the formation of intracellular spicules which are then deposited in the mesoglea rather than by its direct mineralization. That, however, is not the only system of mineral deposition. The corals are capable of extracellular mineralization over large areas of the epidermis (or calicoblastic ectoderm, Fig. 11.1) to form what is technically an exoskeleton. It is here that we first see the effectiveness of a large epithelial sheet as a mineralizing surface.

For these reasons, biomineralization in the coelenterates is an interest-

Table 11.1 Mineralizing Groups of Coelenterata

Organisms		Types of Mineralization
Class Hydrozoa		
Other Hydrocorallina		
Suborder Milleporina	e.g. Stinging or fire corals	Thin epidermal layer over-lying skeleton
Suborder Stylasterina		Thick tissues overlying skeleton
Class Anthozoa		
Subclass Octocorallia		
Order Stolonifera	e.g. Organ-pipe coral	Calcareous tubes or spicules
Order Telestacea		Calcareous spicules
Order Alcyonacea	e.g. Soft corals	Calcareous spicules
Order Coenothecalia	e.g. Blue coral	Massive skeleton
Order Gorgonacea	e.g. Sea whips, sea fans, sea feathers, precious red coral	Calcareous spicules
Order Pennatulacea	e.g. Sea pens, sea pansy	Calcareous spicules
Subclass Zoantharia		
Order Scleractinia	e.g. Stony corals	Heavy external skeleton

ing area of study, with clear examples of both mineralized and non-mineralized (mesogleal) matrices. There are also intracellular and extra-cellular calcification systems, both of which appear to depend on the uptake and transport of calcium from some other region of the animal to the sites of deposition. The most interesting of all the influences in the coelenterates is, however, that of symbiosis. Many of the most effective reef-building corals contain within their tissues symbiotic dinoflagellates or zooxanthellae. The photosynthetic activity of these algae may have a massive influence, enhancing the process of calcification in their hosts by as much as 10-fold. However, the algae are always in the endodermal cells of the coelenterates, some distance away from the calcifying cells of the epidermis. In fact, the fastest growing regions of a coral, at the tips of a branch, are often devoid of zooxanthellae, so that their influence must be a general rather than a localized one.

The implications of these phenomena are very wide ranging and, in order not to confuse them, we will start our discussion with a coelenter-ate that does not contain algae but depends on intracellular spicule formation. Examples of this are found in the Gorgonacea or horny corals of which there are three morphological types: the sea whips (and sea

Figure 11.1   Diagram to show the epithelial cell layers in a coenosteal region of a colonial reef coral. The large round bodies in certain endodermal cells represent zooxanthellae (Johnston, 1980). (Adapted from Vandermeulen, 1972.)

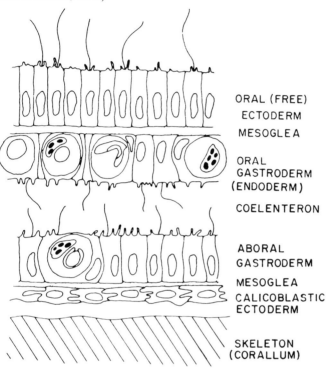

ORAL (FREE) ECTODERM

MESOGLEA

ORAL GASTRODERM (ENDODERM)

COELENTERON

ABORAL GASTRODERM

MESOGLEA

CALICOBLASTIC ECTODERM

SKELETON (CORALLUM)

feathers), the sea fans, and the precious red coral (Table 11.1). The sea whips and sea feathers have branched elongate flexible rods of small diameter; and the sea fans, as their name implies, are of greatly flattened construction. They are built on a plan that permits movement and resistance to wave action and currents (Wainwright and Dillon, 1969; Wainwright and Koehl, 1976). Flexing of the skeleton becomes possible in that it consists largely of an axis of collagenous material with a mineral skeleton of $CaCO_3$, usually in the form of nonarticulated spicules and granules. The central organic material in the red coral *Corallium* is replaced by a solid axis of fused red calcareous spicules that gives a much more rigid structure.

I. Horny Corals—Gorgonacea

A. Structure

The parts of the mineralizing system of the gorgonian *Leptogorgia virgulata* are shown in Fig. 11.2. The outer surface of the branches of a sea whip are covered by an epithelium and by large numbers of feeding polyps. Internally, the central supporting axis has two regions, an inner spongy chambered medulla and a hardened cortex surrounded by an epithelium probably consisting of invaginated epidermal cells that are presumably responsible for secreting it. The axis contains cross-linked collagen, the synthesis of which can be followed by using ^{14}C-labeled

Figure 11.2 Diagram of *Leptogorgia virgulata*. Arrows show the pathways of Ca^{-2} from seawater to deposition in the axis and spicules in cells of the mesoglea. (Courtesy of R. Kingsley, modified.)

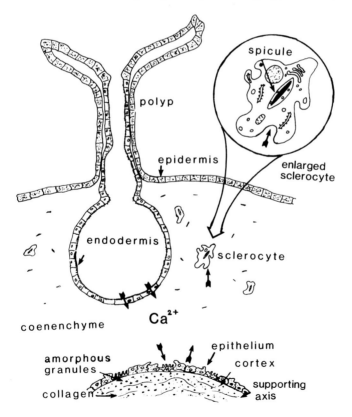

proline (Leversee, 1972, 1980). The axis may be strengthened by the deposition of amorphous granules. Between the outer epithelium and the axis is the coenenchyme, a layer of collagenous mesoglea containing spicule-forming sclerocyte cells. These are ectodermal cells which migrate into the mesoglea and deposit spicules (Fig. 11.2). It is this phenomenon which is our main concern.

Electron diffraction and polarizing microscopy of the spicules show that they have a polycrystalline layered structure with the c-axis of the crystals nearly parallel with the long axis of the spicules (Kingsley and Watabe, 1982). The mineral of the spicules is calcite containing magnesium carbonate (Velimerov and Böhm, 1976) while the matrix consists of a soluble fraction, high in aspartic acid residues (Kingsley and Watabe, 1983), and an insoluble fraction of somewhat similar composition (Watabe *et al.*, 1986).

B. Spicule Formation

The spicules of *L. virgulata* are formed singly within the vacuole of a scleroblast cell (Kingsley and Watabe, 1982). They first appear along with some fibrous material as rod-shaped structures. During development, more fibrous material is added, apparently through fusion of small vacuoles with the spicule-containing vacuole. This fibrous material probably becomes a portion of the organic matrix of the spicule. With lengthening of the spicule, excrescences of mineral are added (Fig. 11.3A). As the spicule becomes still longer, the vacuole membrane becomes stretched by the spicule until it is brought into contact with the plasma membrane. The two membranes then rupture and the spicule passes out of the cell into the mesoglea (Fig. 11.3B). This is an interesting system since sclerocytes in culture also form spicules, but at a markedly reduced rate (Kingsley *et al.*, 1987).

Three aspects of this type of mineralization are of considerable interest. First, the pathways of Ca from the seawater to the sites of mineral deposition within the mesoglea and axis of *L. virgulata* have been followed by pulse–chase experiments and autoradiography using ^{45}Ca (Fig. 11.2, arrows) (Kingsley and Watabe, 1985). Ca from the seawater crosses the epithelium of the polyps and/or the body epithelium and enters the mesoglea. The details of this uptake system are not known. From the mesoglea, Ca moves across the epithelium surrounding the axis and supplies Ca for the amorphous granules there and it also enters the sclerocyte cells to form a spicule. Presumably, the epithelial cells at the axis surface actively transport Ca inward to attain a concentration at

Figure 11.3 (A) Scanning electron micrograph of *Leptogorgia virgulata* spicules. a, Immature spicule with sites of future elaboration (arrow); b, maturing spicule developing "wartlike" projections; c, older spicule with more elaborate "warts"; d, mature spicule with fused "warts"; e, young spicule with more "warts"; f, larger, more developed spicule; g, mature, ornate spicule (Kingsley and Watabe, 1982). (B) Scanning electron micrograph of a spicule being extruded from a scleroblast (Kingsley *et al.*, 1987). (*Figure continues.*)

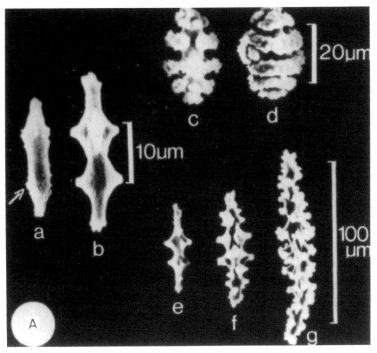

which the granules will precipitate. This, however, seems to be a process that is easily saturated, for once the axis is labeled with ^{45}Ca, calcium can be shown to move outward to the mesoglea and contribute to the source for spicule formation within the sclerocytes. We have, therefore, an epithelial uptake system and a mesogleal transport system supplying calcium for the skeleton of the sea whip. If we now concentrate on the sclerocyte system, then the second aspect of interest is the transport of ions through the cytoplasm into the vacuole until the mineral solubility product is exceeded. Finally, there is the control of crystal orientation and growth which provides the pattern of development from a relatively simple mineral structure of oriented crystals to a more com-

Figure 11.3 (*continued*)

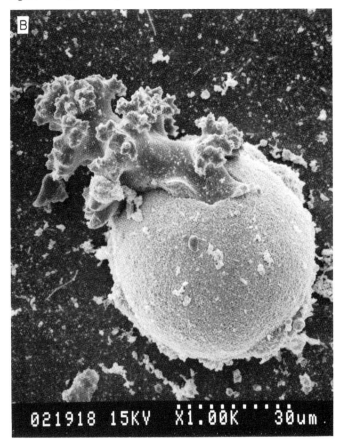

plex construction. This control resides with the organic matrix within the vacuole and/or the vacuole membrane which orients the matrix molecules. The influence of the organic matrix in controlling spicule form is shown by placing decalcified spicule matrix in a calcifying solution (Watabe *et al.*, 1986). The matrix becomes remineralized in a form corresponding to the original crystal form, and in some cases, the remineralized spicule has the normal shape. The mineral content on remineralization is much less than that of normal spicules, but this may not be too surprising since there is probably some loss of soluble matrix during decalcification.

II. Stony Corals—Scleractinia

A. Calcification

What we have described thus far in the coelenterates is a fairly typical system of intracellular calcification, in this case by the sclerocyte cell. To this system we have had to add two new aspects. First, the uptake of calcium by the epidermal cells and second, the transport of ions through the mesoglea. In the scleractinian corals, however, we find another form of biomineralization. The epidermis itself is directly responsible for mineral deposition.

It is possible to appreciate this phenomenon by tracing the formation of the first mineral that is produced when a coelenterate larva, the planula, settles on the substrate. The larva attaches itself by the aboral surface and the ectoderm at this site undergoes a process of cell flattening and organelle change (Vandermeulen, 1974, 1975; Johnson, 1978). At metamorphosis into a polyp this layer will become the squamous calicoblastic ectoderm which forms the calcareous skeleton. The rest of the ectoderm remains as a columnar epidermis secreting mucus and forming an epithelium over the whole organism. Meanwhile, the calicoblastic epithelium is forming the initial deposit of the larval plate, which is sparsely mineralized, consists of small and rather poorly oriented crystals, and contains calcite instead of the entirely aragonitic mineral of the adult. The edges of this basal plate are extended as a crystalline sheet, called the epitheca, that may become the substratum on which the calicoblastic epithelium deposits calcium carbonate to give increased thickening (Barnes, 1972).

The organic matrix of the skeleton is released by the calicoblastic epithelial cells into the subepithelial space where crystals form and grow within the matrix. In larvae of the coral *Ballanophyllia*, Kinchington (1981) thought that the matrix had its origin in intracellular vesicles which disintegrated externally in the subepithelial space. Vesicles of a different appearance were observed within calicoblastic cells, in the intercellular spaces, and at the subepithelial spaces of *Pocillopora damicornis* (Johnston, 1980). These were thought to be the source of organic material enveloping the crystals, but it is not certain that the cells in this immediate region were actually involved in biomineralization. Analyses of matrix following skeletal decalcification have shown the presence of protein, polysaccharides, lipids, and chitin but in very variable amounts (Wilfert and Peters, 1969; Young, 1971). In such analyses, there is always the possibility of inclusions from algae, bacteria, and fungi within the skeleton, but it is generally agreed that the matrix contains a large amount of aspartic acid that could be involved in binding calcium ions (Mitterer, 1978).

The ultrastructure of the adult calicoblastic epithelium shows large intracellular spaces or vesicles and highly interdigitated lateral membranes often opening to the mesoglea but closed on the skeletal side by tight junctions (Fig. 11.1) (Johnston, 1980). Relatively little is known about how the ions reach the sites of mineral deposition. It has been suggested that they pass through the intercellular spaces (i.e., paracellular) or are delivered by the discharge of intracellular vesicles. There is, however, no strong evidence in favor of either route, although it appears unlikely that the crystals reported by Hayes and Goreau (1977) as intracellular precursors of the skeleton are actually involved in biomineralization (Watabe, 1981a,b; Chalker, 1983). One is left, therefore, with a rather unsatisfactory situation. It is generally agreed that ions must be taken up from the seawater by other parts of the coral, and that they must transverse the mesoglea to reach the mineral-depositing calicoblastic cells. Somewhere in the process ATP is used. This would explain why calcification of *Acropora formosa* in the dark is still inhibited by CCCP (carbonyl cyanide-*m*-chlorophenyl hydrazone), an uncoupler of oxidative phosphorylation (Barnes, 1985). The ATP could be used, however, either in the active transport of ions, secretion of matrix, or in the maintenance of cell integrity. It is apparent, therefore, that the cellular processes of biomineralization in the scleractinians are poorly understood, although much more is known about the structure and growth of the skeletal crystals.

B. Crystalline Structures and Growth

The structure of crystal aggregates of corals is complex and unfortunately the nomenclature is confused. The units from which the skeleton is built include spherulites, crystal needles, fusiform crystals, and others (Bryan and Hill, 1941; Sorauf, 1970, 1974; Jell, 1974; Gladfelter, 1983). The crystals are grouped in various arrangements to form increasingly complex structures: needles in very large numbers form fasciculi (Gladfelter, 1983); individual crystals are fused into lathlike units (Sorauf, 1970; Vandermeulen and Watabe, 1973); some crystals are grouped as lamellae that may be joined as trabeculae (Barnes, 1970; Jell, 1974); while others form sclerodermites, which are radial three-dimensional crystal structures originating from a center (Wise and Hay, 1965).

During the growth of a coral, crystals are continually being formed and grouped into the macroscopic patterns characteristic of the species. Within this overall pattern, there are often variations of the type described by Gladfelter (1983) in the growth of the branching coral *Acropora cervicornis*.

Here, the control of crystal form and orientation appears to depend on four factors: the composition of the medium at the crystallization sites, oriented organic matrix molecules, the external surface of the calicoblastic epithelial cells, and a variety of environmental influences. Crystals that are nucleated in contact with the cell surface, or by matrix molecules that are in contact with the cell surface, generally become similarly oriented.

Cellular influences are also seen in the growth of spines at the tips of branches. As the spines grow in length, they appear to penetrate the overlying mesoglea and cell layers. The structure of such a branch of *A. cervicornis* is shown in Fig. 11.4 (Gladfelter, 1982). The wall with spines consists of four mineralized cylinders connected with bars. The cylinders are fenestrated, except for the outer cylinder which has a solid wall. This complex structure is formed by the calicoblastic epithelium that extends throughout the pores and spaces. The individual spines de-

Figure 11.4   Scanning electron microscope views of the tip of a branch of *Acropora cervicornis*. (A) View of axial corallite with tissue removed, showing position of skeletal elements. The wall (WA) is composed of extending spines which connect to one another tangentially by bars (synapticulae) and radially by porous portions of sclerosepta. Projecting into the calyx (CA) are nonperforated portions of the sclerosepta (SE). (B) Enlargement of A showing details of the tip of the axial corallite, and a radially projecting nonperforated unit of the skeleton, the costa (COS). Note the typical scalelike appearance of the surface of the skeleton due to the fasciculi (Gladfelter, 1982).

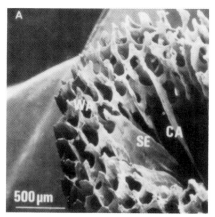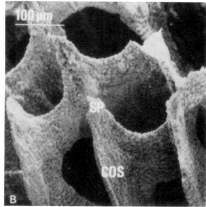

Figure 11.5 Diagrammatic sequence of crystal deposition and growth on an extending spine of the axial corallite. (A) Fusiform crystals deposited in random orientation on the extending spine. Below these occur tufts of crystals, which eventually join to form fasciculi. (B) New fusiform crystals deposited on a spine, resulting in extension of the skeleton. (C) Needlelike crystals grow from the surface of the fusiform crystals in random orientation. If crystal growth is directed toward another crystal, it becomes inhibited. If growth is normal to the surface of the spine, it continues so that tufts of crystals are formed. (D) Several tufts adjacent to one another grow at the same rate and with needles parallel to each other. (E) The tufts join to become fasciculi, forming definitive scalelike fabric of the skeleton (Gladfelter, 1982).

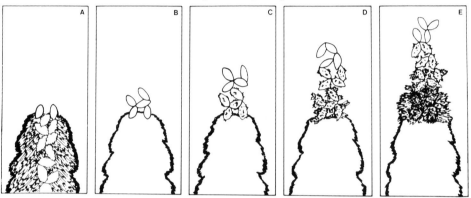

velop by the deposition and growth of randomly oriented crystals with smaller needlelike crystals projecting from them (Fig. 11.5). Slightly proximal to the tip, groups of needlelike crystals project normal to the calicoblastic epithelium and form tufts which then unite forming fasciculi. As the branch ages, these needle-like crystals elongate, filling in the spaces in the scaffolding of the spines.

Environmental conditions that alter the rate of mineral deposition result in changes in crystalline structure. Two examples will illustrate this: (1) During the night, which is a period of less rapid mineralization, fusiform crystals form a framework on distal edges of the spines and bars of *A. cervicornis* (Gladfelter, 1983). At noon, these surfaces become broadened due to the development of initial stages of fasciculi. The changes are correlated with the form of the calicoblastic epithelium. The lappet tissue which forms the epitheca also changes its position depending on the time of day (Barnes, 1972). (2) Tips of spines of wave-exposed colonies of *Pocillopora damicornis* were found to be branched, whereas spines of colonies in sheltered environments were simple cones (Brown *et al.*, 1983). Two aspects of spine growth are noteworthy in these examples. The calicoblastic epithelium does not necessarily deposit crystals uniformly, but calcification for a particular region can be determined by light intensity. The result is a nonuniform deposition rate over a daily cycle. Second, the environment may influence the form of the calicoblastic epithelium in a manner that will alter the morphology of the growing spines.

C. Calcification Rate

Variation in the rate of mineral deposition is characteristic of coral growth. The rate varies at different regions of a single colony (Land *et al.*, 1975), among colonies of the same species, and, as mentioned, with time of day, although this may simply reflect changes in the photosynthetic activity of the zooxanthellae. X-radiographs show growth bands resulting from density differences per unit thickness of the coral (Fig. 11.6); and the bands provide evidence of change in the deposition rate during the course of each year (Buddemeir *et al.*, 1974; Chalker *et al.*, 1985; Wellington and Glynn, 1983). Many factors, environmental and biological, have been considered as possible agents for these variations (Oliver, 1984) but we will concentrate on only one, namely light, since this has the most dramatic effect on hermatypic corals and the greatest theoretical implications.

Figure 11.6 X-radiographic positives showing the changes in skeletal density from January 17 to November 19, 1979, for *Pavona gigantea* at Contadora. Scale bar represents 1.3 cm. The time series samples were collected from the same colony over the time periods indicated. Solid lines connnect the high-density bands formed in the preceding wet, nonupwelling season (June through December). Linear dimensions of low-density (LD) and high-density (HD) bands are shown on the right. Dotted lines on the November 19 sample indicate position of Alizarin Red stain lines. Corals were stained on January 25 (1), May 24 (2), and September 25, 1979 (3). The formation of the HD band corresponds to lower light levels and the production of gametes (Wellington and Glynn, 1983).

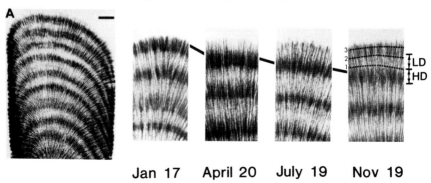

Jan 17   April 20   July 19   Nov 19

Several studies have shown that differences in mineralization rate over daily and annual cycles are related to differences in light intensity, the highest rates occurring during periods of greatest intensity (e.g., Goreau, 1959; Pearse and Muscatine, 1971; Valdermeulen *et al.*, 1972; Chalker *et al.*, 1985). As the intensity of light is increased, the rate of calcification of *A. cervicornis* also increases up to a maximum (Fig. 11.7).

Photosynthesis inhibitors such as DCMU also inhibit the deposition of calcium carbonate in corals (Fig. 11.8; Barnes, 1985), and it is clear that the effects of light must be mediated by the photosynthetic activation of the zooxanthellae in the endodermal cells. While light and photosynthesis are directly related to an enhancement of the mineralization rate, photosynthesis is not a *sine qua non* of mineral deposition. Thus, calcification still occurs in corals that are kept in darkness (although at a much reduced rate) and ahermatypic corals that lack zooxanthellae also deposit skeletons.

Figure 11.7 Light-saturation curve for light-enhanced calcification by 2-cm tips of *Acropora cervicornis* incubated with solar irradiance for 2 hr about local solar noon. Mean ± one standard error for four individuals. The data are fitted to the function $C = Cm \tanh(I/I_k)$. $Cm = 2.34 \mu mol Ca^{2+} hr^{-1} (2$-cm tip$)^{-1}$. I_k equals 1.01 Einsteins $m^{-2} hr^{-1}$. The filled circle represents corals that are presumed to be photoinhibited. These data were not used in curve fitting (Chalker, 1983).

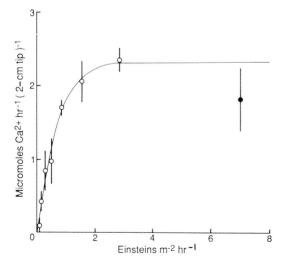

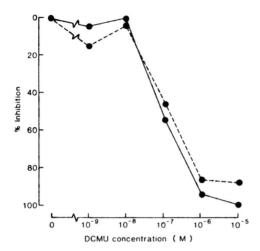

Figure 11.8 Effects of DCMU on gross photosynthesis (solid line) and calcification (dotted line) in *Acropora formosa* (Barnes, 1985).

Data of this type force one to ask what is the metabolic link between the activity of the photosynthesizing dinoflagellates and the calcifying ability of the host—especially since the location of the zooxanthellae and the sites of calcification may be separated by quite large distances in the coral.

There have been two types of hypothesis. The first suggests that the photosynthetic process removes inhibitory materials from the site of calcification. Such materials include (a) carbon dioxide, the removal of which would favor carbonate production (Goreau, 1963); (b) metabolic "wastes" which inhibit general metabolism (Yonge, 1968); and (c) phosphates, which could act as crystal poisons (Simkiss, 1964).

The second set of suggestions emphasize the value of the photosynthetic products and how such carbon sources may stimulate the metabolism of the coral. Such proposals include (d) enhancing metabolic activity, since any increase in ATP synthesis would stimulate a whole range of activities including ion transport (Chapman, 1974; Chalker and Taylor, 1975); and (e) facilitating matrix synthesis (Vandermeulen and Muscatine, 1974).

Early studies suggested that the metabolic influence was glycerol, which was transferred from the photosynthesizing algae to the coral. Subsequent work showed that, although this was a "short-term" product, it was rapidly metabolized so that it may be lipid that is eventually accumulated by the host (Gladfelter, 1985). The overall effect is, however, dramatic and Muscatine *et al.* (1984) have shown that up to 95% of photosynthetically fixed organic material may be translocated from the algae to the coral tissues either as glucose, amino acids, or lipids.

Thus, although the enhancement of calcification by zooxanthellae and photosynthesis is generally accepted, the precise details are poorly understood. It may, therefore, be as well to emphasize that corals can calcify in the dark, that all coelenterates that contain zooxanthellae do not have skeletons (e.g., *Hydra*), and that all coelenterates that form mineralized skeletons do not have zooxanthellae. Perhaps to restore the balance of hypotheses one should simply suggest that excess metabolites are used up by driving the corals' biomineralization process faster.

III. Coral Reefs

The most spectacular result of coelenterate mineralization is, of course, the coral reef. The scleractinian organisms that are largely responsible do not survive in water cooler than about 18°C. This, and the enhancement caused by symbiont photosynthesis, means that coral reefs are largely restricted to shallow waters throughout the South Pacific, in the East Indies and the Indian Ocean to Sri Lanka, around Madagascar along the eastern coast of Brazil, through the West Indies, and off the Florida coast. It is estimated that it requires from 1000 to 30,000 years to form a reef 50 m thick and the most spectacular, the Great Barrier Reef of Australia, is more than 2000 km long.

Coral reefs are, of course, whole ecosystems and have been looked at in those terms. Estimates of the amount of carbon fixed by photosynthesis (P) and the amount released by respiration (R) give values of about 2–6 g carbon m^{-2} day^{-1} for both processes with a P/R ratio of about 1. Different parts of the reef system may, however, show very different rates of inorganic gain (G) but they are typically measured in kg CaCO$_3$ m^{-2} yr^{-1}. The high activity regions (100% hard substrate) are very variable with respect to their location in the reef complex but generally have very high levels of photosynthetic activity ($P = 20$, $G = 10$). More typical of the reef as a whole are the flat algal-encrusted pavements ($P = 5$, $G = 4$) while the shallow lagoon environments showed little activity ($P = 1$, $G = 0.5$) (Kinsey, 1985). Typically, however, a coral reef deposits about 4 kg CaCO$_3$ m^{-2} yr^{-1} (Table 11.2). This means that each year each polyp effectively precipitates all the calcium carbonate out of a column of sea water 4 m high.

The coral reef as a whole appears to be an "exporter" of nitrate, which emphasizes the importance of the algal and microbial components in

Table 11.2 Typical Values for Metabolism of Coral Reef Systems[a]

	Photosynthesis (P) (g carbon m^{-2} day^{-1})	Respiration (R) (g carbon m^{-2} day^{-1})	Inorganic gain (G) (kg $CaCO_3$ m^{-2} yr^{-1})
High activity areas (Abrolhos)	17	17	12
Algal pavement zone (One Tree Island)	2	0.5	4
Reef—flat coral/algae (Enewetak)	6	6	4
Shallow lagoon (One Tree Island)	3.1	3.1	1.5

[a] After Kinsey (1985), from various sources.

nitrogen fixation. Phosphorus is a more limited nutrient and is largely recycled within the reef ecosystem. When these estimates of reef activity are related to the biochemical events of the coral mineralization systems, some additional features become apparent. According to Muscatine *et al.* (1984), the growth rate of the zooxanthellae is very low and they account for less than 2% of the fixed organic carbon. The remaining carbon is translocated to the coral but much of it is then released as dissolved organic matter. In fact, up to 40% of the carbon fixed by photosynthesis is released as mucus (Gladfelter, 1985). Calcification is the most dramatic product of the corals' activities but it represents only a minor component of the net flux of metabolites through this system.

IV. Summary

1. Coelenterates possess a simple structure of two epithelia separated by a glycoprotein matrix, the mesoglea. This matrix does not calcify directly but it may be strengthened by intracellular spicules to form such structures as the axial skeleton of the sea whips.
2. A second system of mineral deposition is associated with specializations in the ectoderm to form the calicoblastic epidermis. This epithelial system is capable of producing large quantities of calcium carbonate to form the massive deposits of the stony corals.
3. The supply of calcium to these sites of mineralization involves translocation to various parts of the organism. Proteinaceous material, secreted by the calicoblastic epidermis and present within mineralizing vacuoles, becomes the matrix of the mineral deposits. *In vitro*

studies on the organic material of gorgonian spicules show that it can be remineralized and that the matrix influences the form of the mineral.

4. The complex mineral structure of hermatypic corals is partly dictated by the calicoblastic epithelium.

5. The possession of photosynthesizing zooxanthellae within the endoderm of hermatypic corals increases the rate of calcium carbonate deposition. The rate of calcification depends on light intensity. Several hypotheses have been advanced to explain this but the phenomenon is not completely understood. Light is not a requirement for growth and mineral deposition in corals as evident from measurements of hermatypic corals in the absence of light and growth of ahermatypic corals which lack photosynthetic symbionts.

6. The success of corals as reef builders is the result of myriad interconnected polyps which, by their individual deposition of $CaCO_3$, contribute to a single large skeleton. Calcifying algae play an important role in consolidating reef structure.

References

Barnes, D. J. (1970). Coral skeletons: An explanation of their growth and structure. *Science* **170**, 1305–1308.

Barnes, D. J. (1972). The structure and function of growth ridges in scleractinian coral skeletons. *Proc. R. Soc. London, Ser. B* **182**, 331–350.

Barnes, D. J. (1985). The effect of photosynthetic and respiratory inhibitors upon calcification of the staghorn coral, *Acropora formosa. Proc. Int. Coral Reef Congr.* 23.

Brown, B. E., Hewit, R., and Le Tissier, M. D. (1983). The nature and construction of skeletal spines in *Pocillopora damicornis* (Linnaeus). *Coral Reefs* **2**, 81–89.

Bryan, W. H., and Hill, D. (1941). Spherulitic crystallization as a mechanism of skeletal growth in the hexacorals. *Proc. R. Soc. Queensl.* **52**, 78–91.

Buddemeier, R. W., Margos, J. E., and Knutson, D. K. (1974). Radiographic studies of reef coral exoskeletons: Rates and patterns of coral growth. *J. Exp. Mar. Biol. Ecol.* **44**, 179–200.

Chalker, B. E. (1983). Calcification by corals and other animals on the reef. *In* "Perspectives on Coral Reefs" (D. J. Barnes, ed.), pp. 29–45. Clouston, Manuka, Australia.

Chalker, B. E., and Taylor, D. L. (1975). Light-enhanced calcification and the role of oxidative phosphorylation in calcification of the coral *Acropora cervicornis. Proc. R. Soc. London, Ser. B* **190**, 323–331.

Chalker, B., Barnes, D., and Isdale, P. (1985). Calibration of x-ray densitometry for the measurement of coral skeletal density. *Coral Reefs* **4**, 95–100.

Chapman, G. (1974). The skeletal system. *In* "Coeleterate Biology: Reviews and New Perspective" (L. Muscatine and H. M. Lenhoff, eds.), pp. 93–128. Academic Press, New York.

Gladfelter, E. H. (1982). Skeletal development in *Acropora cervicornis*. I. Patterns of calcium carbonate deposition in the axial corallite. *Coral Reefs* **1**, 45–51.

Gladfelter, E. H. (1983). Skeletal development in *Acropora cervicornis*. II. Diel patterns of calcium carbonate action. *Coral Reefs* **2**, 91–100.

Gladfelter, E. H. (1985). Metabolism, calcification and carbon production. II. Organism-level studies. *Proc. Int. Coral Reef Congr., 5th* **4**, 527–539.

Goreau, T. F. (1959). The physiology of skeleton formation in corals. I. A method for measuring the rate of calcium deposition by corals under different conditions. *Biol. Bull. (Woods Hole, Mass.)* **116**, 59–75.

Goreau, T. F. (1963). Calcium carbonate deposition by coralline algae and corals in relation to their roles as reef-builders. *Ann. N.Y. Acad. Sci.* **109**, 127–167.

Hayes, R. L., and Goreau, N. I. (1977). Intracellular crystal-bearing vesicles in the epidermis of scleractinian corals, *Astrangia dance* (Agassiz) and *Porites porites* (Pallas). *Biol. Bull. (Woods Hole, Mass.)* **152**, 26–40.

Jell, J. S. (1974). The microstructure of some scleractinian corals. *Proc. Int. Coral Reef Symp., 2nd* **2**, 301–320.

Johnson, M. F. (1978). A comparative study of the external form and skeleton of the calareous sponges *Clathrina coriacea* and *Clothrina blanca* from Santa Catalina Island, California. *Can. J. Zool.* **56**, 1669–1677.

Johnston, I. S. (1980). The ultrastructure of skeletogenesis in hermatypic corals. *Int. Rev. Cytol.* **67**, 171–214.

Kinchington, D. (1981). Organic matrix synthesis by scleractinian coral larval and post-larval stages during skeletogenesis. *Proc. Int. Coral Reef Symp., 4th* **2**, 107–113.

Kingsley, R. J., and Watabe, N. (1982). Ultrastructure of the axial region in *Leptogorgia virgulata* (Cnidaria: Gorgonacea). *Trans. Am. Microsc. Soc.* **101**, 325–339.

Kingsley, R. J., and Watabe, N. (1983). Analysis of proteinaceous components of the organic matrices of spicules from the gorgonian *Leptogorgia virgulata*. *Comp. Biochem. Physiol. B* **76**, 443–447.

Kingsley, R. J., and Watabe, N. (1985). An autoradiographic study of calcium transport in spicule formation in the gorgonian *Leptogorgia virgulata* (Lamarck) (Coelenterata: Gorgonacea). *Cell Tissue Res.* **239**, 305–310.

Kingsley, R. J., Bernhardt, A. M., Wilbur, K. M., and Watabe, N. (1987). Scleroblast cultures from the gorgonian *Leptogorgia virgulata* (Lamarck) (Coelenterata: Gorgonacea). *In Vitro Cell Dev. Biol.* **23**, 297–302.

Kinsey, D. W. (1985). Metabolism, calcification and carbon production. I. Systems level studies. *Proc. Int. Coral Reef Congr., 5th* **4**, 505–526.

Land, L. S., Lang, J. C., and Barnes, D. J. (1975). Extension rate: A primary control on the isotopic composition of West Indian (Jamaican) scleractinian reef coral skeletons. *Mar. Biol.* **33**, 221–223.

Leversee, G. J. (1972). "Organization and Sythesis of the Axial Skeleton of *Leptogorgia virgulata*," Ph.D. thesis. Duke University, Durham, North Carolina.

Leversee, G. J. (1980). Incorporation and distribution of labelled proline in collagenous and non-collagenous components of the gorgonian coral *Leptogorgia virgulata* (Coelenterata, Octocorallia). *Comp. Biochem. Physiol. B* **67**, 499–503.

Mitterer, R. M. (1978). Amino acid composition and metal binding capability of the skeletal protein of coral. *Bull. Mar. Sci.* **28**, 171–180.

Muscatine, L., Falkowski, P. G., Porter, J. W., and Dubinsky, Z. (1984). Fate of photosynthetic carbon in light and shade adapted colonies of the symbiotic coral *Stylophora pistillata*. *Proc. R. Soc. London, Ser. B.* **222**, 181–202.

Oliver, J. K. (1984). Intra-colony variation in the growth of *Acropora formosa:* Extension rates and skeletal structure of white zooxanthellae-free and brown-tipped branches. *Coral Reefs* **3**, 139–147.

Pearse, V. B., and Muscatine, L. (1971). Role of symbiotic algae (zooxanthellae) in coral calcification. *Biol. Bull. (Woods Hole, Mass.)* **141**, 350–363.

Simkiss, K. (1964). Phosphates as crystal poisons of calcification. *Biol. Rev.* **39**, 487–505.

Sorauf, J. D. (1970). Microstructure and formation of dissepiments in the skeleton of the recent Scleractinia (hexacorals). *Biomineralisation* **2**, 2–22.

Sorauf, J. E. (1974). Observations on microstructure and biocrystallization in Coelenterates. *Biomineralisation* **7**, 37–55.

Vandermeulen, J. H. (1972). "Studies on Skeleton Formation, Tissue Ultrastructure, and Physiology of Calcification in the Reef Coral *Pocillopora damicornis* Lamark. Ph.D. dissertation. Univ. of California at Los Angeles.

Vandermeulen, J. H. (1974). Studies on reef corals. II: Fine structure of planktonic planula larva of *P. damicornis*, with emphasis on the aboral epidermis. *Mar. Biol.* **27**, 239–249.

Vandermeulen, J. H. (1975). Studies on reef corals. III: Fine structural changes of calicoblast cells in *Pocillopora damicornis* during settling and calcification. *Mar. Biol.* **31**, 69–78.

Vandermeulen, J. H., and Muscatine, L. (1974). Influence of symbiotic algae on calcification in reef corals: Critique and progress report. *In* "Symbiosis in the Sea" (W. Vernberg, ed.), pp. 1–19. Univ. of South Carolina Press, Columbia, South Carolina.

Vandermeulen, J. H., and Watabe, N. (1973). Studies on reef corals. I. Skeleton formation by newly settled planula larva on *Pocillopora damicornis*. *Mar. Biol.* **23**, 47–57.

Vandermeulen, J. H., Davis, N., and Muscatine, L. (1972). The effect of inhibitors of photosynthesis on zooxanthellae in corals and other marine invertebrates. *Mar. Biol.* **16**, 185–191.

Velimerov, B., and Böhm, E. L. (1976). Calcium and magnesium carbonate concentrations in different growth regions of gorgonians. *Mar. Biol.* **35**, 269–275.

Wainwright, S. A., and Dillon, J. R. (1969). Orientation of sea fans. *Biol. Bull. (Woods Hole, Mass.)* **136**, 130–139.

Wainwright, S. A., and Koehl, M. A. R. (1976). The nature of flow and the reaction of benthic Cnidaria to it. *In* "Coelenterate Ecology and Behavior" (G. O. Mackle, ed.), pp. 5–21. Plenum, New York.

Watabe, N. (1981a). Some problems on the structure and formation of calcium carbonate crystals and their aggregates in the invertebrates. *In* "Study of Molluscan Paleobiology" (T. Habe and M. Omori, eds.), pp. 34–36. Univ. of Niigata, Niigata, Japan.

Watabe, N. (1981b). Crystal growth of calcium carbonate in the invertebrates. *Prog. Cryst. Growth Charact.* **4**, 99–147.

Watabe, N., Bernhardt, A. M., Kingsley, R. J., and Wilbur, K. M. (1986). Recalcification of decalcified spicule matrices of the gorgonian *Leptogorgia virgulata* (Cnidaria: Anthozoa). *Trans. Am. Microsc. Soc.* **105**, 311–318.

Wellington, G. M., and Glynn, P. W. (1983). Environmental influences on skeletal banding in eastern Pacific (Panama) corals. *Coral Reefs* **1**, 215–222.

Wilfert, M., and Peters, W. (1969). Vorkommen von Chitin bei Coelenteraten. *Z. Morphol. Tiere* **64**, 77–84.

Wise, S. W., and Hay, W. W. (1965). Ultrastructure of the septa of scleractinian corals. *Spec. Pap. Geol. Soc. Am.* **87**, 187–188.

Yonge, C. M. (1968). Living corals. *Proc. R. Soc. London, Ser. B* **169**, 329–344.

Young, S. C. (1971). Organic material from scleractinian coral skeletons. I. Variation in composition between several species. *Comp. Biochem. Physiol. B* **40**, 113–120.

12 | Annelids—Glandular Secretions

As builders of mineral structures, the annelids are a modest phylum. Their constructions are invariably simple and fall short of the elegance and complexity of forms seen in protozoan, echinoderm, and molluscan skeletons. The simplest structures are merely granular aggregates of calcium carbonate, formed in the esophageal glands of the lumbricid earthworms and, although the serpulid worms form a calcareous tube, it too is derived from a granular secretion that is compacted and molded into a sleeve encasing the body. Mineral deposition by annelids has received more descriptive than experimental attention, but it is of interest to us here because it illustrates how an epithelial system of mineral deposition can be elaborated into glandular structures that meet the particular needs of these organisms.

Our discussion will consider mineral deposition in two families: the Lumbricidae with their simple type of glandular mineralization and the adult Serpulidae which are involved in tube secretion.

I. Lumbricidae

The Lumbricidae, a family of the oligochaete annelids, includes some 220 species (Satchell, 1967). Familiar examples that have received attention experimentally are the earthworms *Lumbricus terrestris*, *Allolobophora calignosa*, and *Eisenia foetida*. In contrast to the polychaetes, the Lumbricidae do not form a calcified tube, but they have important influences on the physical and chemical properties of the soil, including the addition of calcareous concretions (Lee, 1985).

A. The Mineralizing System

The formation of $CaCO_3$ by earthworms takes place in pouchlike diverticula called calciferous glands that lie along the esophagus (Fig. 12.1). They occur in one or more pairs in segments 10 to 14 and their formation is described in detail in Stephenson's (1930) monograph on the Oligochaeta. Initially, the esophageal epithelium is simply elaborated into prominent folds but these frequently fuse to form a series of longitudinal tunnels. The walls of these folds are highly vascularized and where they fuse they enclose a series of large blood sinuses running between the single-celled layers of epithelial cells (Fig. 12.1). As we shall see, this

Figure 12.1 Transverse section of calciferous glands of earthworm to show disposition of gut epithelium, blood sinuses, and chloragogen tissue (Laverack, 1963). (Redrawn from Gabbay, 1958.)

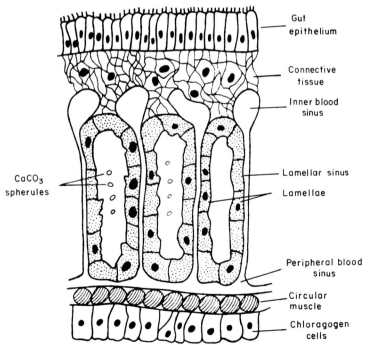

meets two crucial needs of a biomineralizing system. By elaborating the epithelium into a gland, it encloses a microenvironment for mineral deposition, while the large blood sinuses provide a ready supply of ions.

In *Lumbricus* the first calcareous bodies that form within the glands are 0.75–5 μm in diameter. Within the pouches the concretions resulting from the coalescence of these deposits are usually 0.5–1 mm in size and consist of spherites and crystalline blocks, which presumably result from a phase change of spherites (Morgan, 1981). The mineral is usually calcite (Robertson, 1936; Crang *et al.*, 1968) but in *Eisenia foetida* no X-ray diffraction pattern could be obtained, indicating amorphous calcium carbonate (van Gansen-Semal, 1959). Organic matter amounting to 3–5% is scattered throughout the concretions in *Lumbricus* (Robertson, 1936).

The sources of the calcium are ingested plant material, soil, and mineral that enter the gut, and a small amount of calcium that enters through the skin (Piearce, 1972). Calcium ions move across the gut wall and into the blood that carries them forward to the calciferous glands where the formation of concretions of $CaCO_3$ takes place. The concretions pass from the glands to the lumen of the esophagus, then through the gut and to the soil. Thus, calcium is transported from the gut and back to the gut via the blood and calciferous glands.

B. Mineralizing Mechanisms

1. Intake of Ions   The rate of movement of calcium into the blood, to the epithelial cells of the calciferous glands, and finally into the calcareous concretions has been followed by autoradiography using ^{45}Ca (Nakahara and Bevelander, 1969). Thirty minutes after the injection of the isotope into the coelomic cavity, the ^{45}Ca was present within the epithelial cells; after 60 min, ^{45}Ca was still within the cells but was also present in the concretions within the gland lumina; and after 4 hr, the isotope was confined almost entirely to the concretions.

Worms which ingest carbonate can solubilize at least part of this to calcium and bicarbonate ions in the gut and convert the ions to calcium carbonate in the calciferous glands. Worms fed calcium salts other than calcium carbonate also precipitate carbonate, indicating its respiratory origin from carbon dioxide and bicarbonate ions. By measuring both the amount of carbonate precipitated and the respiratory rate over a given interval, Robertson (1936) found that on average only 5.3% of the respiratory CO_2 was required to account for the carbonate precipitated.

2. Mineralization by the Calciferous Glands The structure of the glands is especially favorable for relatively rapid ion movement into them. As shown in Fig. 12.2, the peripheral blood sinuses are immediately contiguous to the lamellar cells with the result that the path length of ion transport will be the thickness of the blood sinus wall and a single epithelial cell. In *Eisenia* the basal and lateral regions of the cells have deeply infolded membranes (Fig. 12.2), a type of structure commonly associated with extensive ion transport (Jamieson, 1981). Once calcium passes from the blood sinus through this membrane into the cytosol, a portion of it is transported into vacuoles in which spherules of calcium carbonate are formed (van Gansen-Semal, 1959; Crang *et al.*, 1968). These bodies are then released into the lumen of the gland by the breakdown of the apical cell membrane (van Gansen-Semal, 1959; Crang *et al.*, 1968).

A second advantageous feature of the calciferous glands is the small diameter of the lumina, each completely bounded by epithelial cells. This provides a relatively large surface-to-volume ratio, facilitating the movement of ions and the precipitation of calcium carbonate. Spheruliths are formed at this site (Laverack, 1963; Jamieson, 1981), indicating that the epithelial cells actively transport ions into the lumen.

The small calcareous bodies within the lumina form a milky suspension that passes through the tunnels to the esophogeal pouches. These

Figure 12.2 Schematic representation of the ultrastructure of the calciferous epithelium of *Eisenia foetida* (Jamieson, 1981). (After Chapron, 1971.)

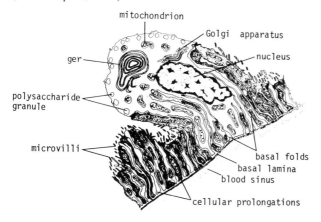

bodies coalesce as concretions and enter the esophagus (Robertson, 1936).

Here, then, within the calciferous glands, two major mineralizing events take place. One is a unidirectional movement of ions across the cell membrane adjoining the blood sinus, through the cytosol, and out through the apical cell membrane to the gland lumen where calcareous bodies are formed. The other is the formation of calcium carbonate deposits within intracellular vacuoles. This mixture of extracellular and intracellular mineralization systems is also of common occurrence in other phyla.

C. Functions of the Calciferous Glands

A number of functions have been proposed for the calciferous glands. These have been summarized and evaluated by Robertson (1936) and Laverack (1963). Of these, only two appear to be important in the economy of the animal. The first is the removal of excess calcium, a function to which Darwin (1881) called attention. This appears to be determined by the amount of calcium entering the blood rather than the amount ingested, although as one might expect, there is a good correlation between the calcium content of the diet and the development of the calciferous glands (Piearce, 1972). The second function of the glands is the buffering of body fluids by calcium carbonate against decreases in pH due to high environmental carbon dioxide. The importance of this function is more difficult to assess under environmental conditions since it depends on both the concentration of carbon dioxide and the time spent in the burrows. The glands do contain the enzyme carbonic anhydrase (Clark, 1957), which would facilitate the release of calcium bicarbonate under these conditions. In these circumstances of hypercapnia, the pH of the coelomic fluid was found to decrease less in normal animals than in those from which the glands had been removed (Dotterweich, 1933). However, it should be noted that species living in anaerobic environments may possess calciferous glands that are relatively inactive (Piearce, 1972).

D. Turnover Rate

The rate at which earthworms take in ingested calcium and effect its turnover as calcium carbonate may be considerable. By collecting calcareous concretions, Robertson (1936) found that *Lumbricus* converted 2–3 mg $CaCO_3$/g body wt/wk. A rough estimate for a population density of 10^6 earthworms per hectare indicates a rate of calcium carbonate formation of the order of 1 kg/ha/yr (Wiecek and Messenger, 1972).

The ecological importance of earthworms to the soil was emphasized by Darwin (1881) and has been discussed in detail by Lee (1985). A significant factor is the fact that they do not build a skeleton. In this, they are unique among mineralizing invertebrates. If their activity had resulted in the construction and increase in mass of a tube that encased the animal as it increased in size, then trituration of the soil through body movements would not have been possible. Moreover, the conversion of ingested food and soil into calcareous concretions would have been restricted, with a decline in this function that is so important for soil fertilization. In the high rate of production of calcareous concretions by these annelids, one is reminded of the coccolithophorid algae that also continuously form and extrude calcareous coccoliths from ions in seawater (see Chapter 6, II).

II. Serpulid Polychaetes

The polychaetes are divided into two main groups, the Errantia and the Sedentaria. The Errantia are the free-moving species, although many of them form tubes. The Sedentaria, as the name indicates, are restricted in their movements and live permanently in tubes or burrows. Of this group, the serpulids are the only annelids that form tubes of calcium carbonate. The tubes are attached to various types of substrata including algae, bryozoans, corals, shells, rocks, buoys, hulls of ships, and water conduits (Day, 1967). The tubes may be straight, curved, or in the form of a flattened spiral (Fig. 12.3a). In cross section, the tubes are usually incomplete in the area of attachment but attached complete tubes are occasionally found. If the tube grows away from the substratum, it is round and complete (Fig. 12.3b; Hedley, 1958).

A. The Mineralizing System

When the free-swimming larva of a polychaete worm such as *Spirorbis spirorbis* settles, it forms a thin anchoring tube around the posterior half of the body. Within the next 3 hr it builds a comparatively thick calcareous tube which surrounds the anterior end of the worm. It does this by unfolding an already-formed collar which is the site of calcification (Nott and Parkes, 1975). There are, in fact, two glandular regions beneath the collar in the anterior portion of the worm and it is from here that the

Figure 12.3 (a) Tubes of *Eupomatus* sp. (Courtesy of Dr. E. Muzii.) (b) Top, left to right: Diagram of an attached portion of tube which is incomplete in cross section; diagram of an attached portion of tube which is complete in cross section; diagram of a portion of tube which has grown away from the substraum and which is completely round in cross section. Bottom: Diagram of the anterior end of the tube with the broken line indicating the region covered by the collar of the worm (drawn on one side of the tube only). Calcareous material is deposited first in the regions A–B and C–D. Further deposits appear on the areas 1, 2, 3, and 4, and on region B–C (Hedley, 1958).

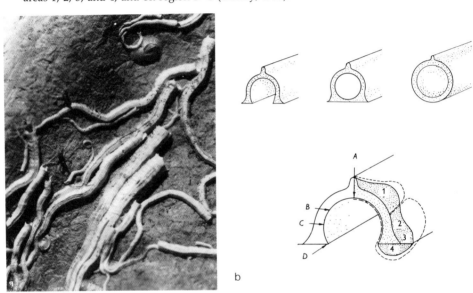

b

material of the tube is secreted (Fig. 12.4). In most polychaetes it is the region in the ventral peristomium in which a pair of calcium-secreting glands is located. The other glandular area is the epithelium of the ventral shield surrounding the openings of the calcium-secreting glands (Fig. 12.4). In addition to the calcium carbonate formed by these two areas, hydroxyapatite and calcium magnesium phosphate crystals are present in cells of *Pomatoceros caeruleus* in two small patches near the junction of the collar and the branchial crown, but these do not appear to be involved in tube formation (Fig. 12.4) (Neff, 1971). The serpulids also have the capacity to repair the posterior end of the calcified tube by the deposition of calcium carbonate. The cells responsible are thought to be the ventral epithelium of that region with possible participation by the rectum (Hedley, 1958).

The mineral of the tube may be aragonite, calcite, or aragonite and

Figure 12.4   Diagram of *Pomatoceros caeruleus* in the position
it assumes in its tube while depositing mineral. The three
areas which deposit mineral are shown: CSG, calcium-secreting glands; VS, ventral shield epithelium; and AC, hy-
droxyapatite-containing cells. The other structures are BR,
branchial crown; C, collar; and T, calcified tube (Neff, 1971).

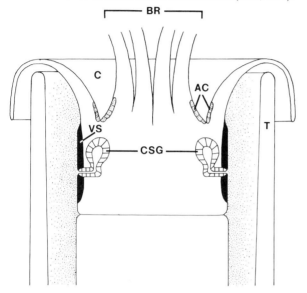

calcite. When both polymorphs are present, the individual polymorphs
occur in separate layers as they do in *Eupomatus dianthus* (Muzii, 1965)
and in molluscan shells. The presence of the two polymorphs in *Eupomatus* appears to be related to their glandular origin, the aragonite being
secreted by the calcium-secreting glands and the high magnesium calcite
coming from the ventral shield epithelium (Neff, 1967).

An examination of the microscopic structure of the calcium-secreting
glands of serpulids has shown considerable variation among species
(Hedley, 1956; Vovelle, 1956; Neff, 1967, 1971). The glands may be made
up of one or more cell types. For example, the outer cell layer of *E.
dianthus* has a mucous-type cell that may well supply a portion of the
acid mucopolysaccharide present in the tube (Fig. 12.5) (Neff, 1967). In
P. caeruleus, a corresponding cell and acid mucopolysaccharide secretion
are not present, and the equivalent function is apparently performed by
the ventral shield epithelium surrounding the gland opening. The second cell type in *Eupomatus* is characterized by a very well-developed

Figure 12.5 A schematic representation of the arrangement of the three types of secretory cells of the calcium-secreting gland of *Eupomatus dianthus*. Cell organelles in the different types of secretory cells are illustrated: F, fiber bundle; G, Golgi complex; GI, glycogen, Lum, lumen; M, mitochondrion; MSV, mucopolysaccharide vacuole; MV, microvilli; MVB, multivesicular body; N, nucleus; R, ribosomes; Sec, secretory products; SV, secretory vacuole (Neff, 1967).

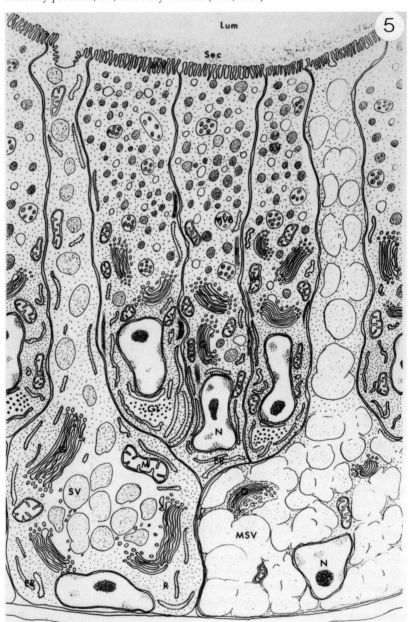

Golgi complex with associated small secretory vesicles and multivesicular bodies.

The major products of these glands are calcareous granules and the matrix in which the granules are suspended (Fig. 12.6). In the case of *Pomatoceros*, the individual granules appear in sections to be squares or rhomboids 150–200 μm in size. Electron diffraction patterns show them to be polycrystalline calcite; and they are composed of needlelike crystallites packed in parallel. Within granules that are incompletely calcified it is possible to see fibrous material with the same orientation as the crystallites. This correspondence suggests some control of crystallites by the organic matrix. When this suspension of granules reaches the external surface of the animal it solidifies by a process that is sufficiently slow to allow the undersurface of the collar (Fig. 12.4) to act as a mold in shaping the end of the tube. This appears to invoke two new phenomena that are more usually associated with the building industry, namely, the solidification of a previously prepared set of granules and the controlled setting of this material.

B. Cellular Mechanisms

The source of calcium for the tubes of serpulid worms is supplied at least in part from the surrounding water. It is therefore reasonable to expect differences in this system between marine, brackish, and freshwater species (Oglesby, 1978). The mechanism by which the calcium of the calcium-secreting gland cells is moved across the cell membrane to the gland lumen is uncertain, with various claims for active transport or exocytosis of granules formed within intracellular vacuoles. Active transport certainly appears likely in some species (see Neff, 1967; Nott and Parkes, 1975); while electron microscopy of the cells of *P. caeruleus* indicates that granules are formed on lamellar organic matrix within multivesicular bodies (Neff, 1967).

If the calcium is moved across the membrane by active transport, the presence of granules in the gland lumen and their increase in size suggest that there is a continual supply of calcium and carbonate ions. Despite this, the granules remain largely separate within the lumen until they reach the external surface of the animal. This is a phenomenon of considerable interest that differs from biomineralization in many other phyla. The maintenance of separated granules indicates that there is either an inhibition of their growth by substances present within the gland lumen or a rate of supply of calcium that will not permit faster growth. Once secreted, the properties of this material rapidly change.

Figure 12.6 The acinar lumen of an active calcium-secreting gland of *Pomatoceros caeruleus* showing calcified secretory granules (Sec G) in some of which an orderly lamellar substructure can be seen (arrows). MV, microvillus. Inset is a limited area electron diffraction pattern from calcified secretory granules indicating that they contain calcite. Cacodylate-buffered glutaraldehyde molybdate-osmium fixation. Unstained. ×40,950 (Neff, 1967).

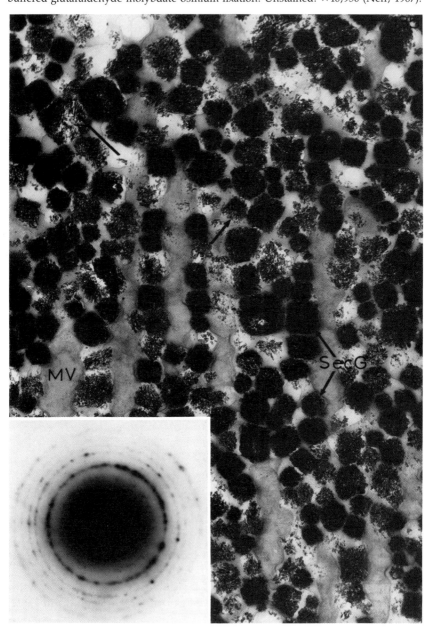

Thus, the value of maintaining the mineral in a granular form is readily apparent since it enables the production of a continually lengthening tube throughout the animal's growth.

It appears that the acid mucopolysaccharide matrix of the tube becomes solidified on contact with seawater. When the tube is decalcified, the tubular form is maintained (Bernhardt et al., 1985). Furthermore, the matrix becomes remineralized on exposure to a solution containing calcium and bicarbonate ions, raising the possibility that seawater may contribute calcium carbonate to the tube at the time of its solidification. It could be concluded, therefore, that the elaboration of an epithelium into a secretory gland not only provides an ideal situation for mineral deposition but it also provides opportunities for the evolution of binding and setting agents that are able to turn granular secretions into consolidated structures.

We stated earlier that the fine structure of a biomineral is determined primarily by the cells directly responsible for its formation. This is dem-

Figure 12.7 Diagrammatic representation of a median–longitudinal section through the opercular filament of *Pomatoceros lamarckii*. (After Bubel, 1983.)

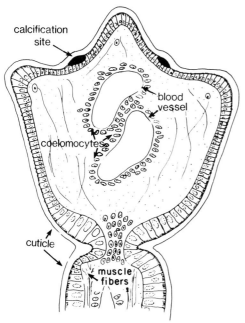

onstrated once again in the serpulids. Whereas the mineral tube is largely lacking in the orientation of its fine structure, the opercular plate at the anterior end of the animal has a structure that is oriented in both its mineral and matrix components. The plate is three-layered, consisting of an outer uncalcified cuticle and two calcified layers with oriented fibrous matrix (Fig. 12.7) (Bubel, 1983). A single layer of epithelium forms the three layers. The difference in structure between the tube and the plate results from the cells responsible and the nature of the microenvironments. The plate is constructed in a nonflowing closed microenvironment between secreting cells and a cuticle that excludes the outer medium. The result is the organization of the organic matrix by self-assembly and the resulting control of crystal orientation.

III. Summary

1. The annelids have a mineralization system that produces granules of calcium carbonate, and in this respect this phylum differs from others. Relatively little experimental attention has been given to this system and many details are lacking.
2. The granules are produced by glands and contain organic material, but detailed analyses have not been carried out. Nor do we have information as to the possible role of this material in crystal nucleation.
3. In the Lumbricidae, the formation of calcareous granules and their excretion provide a method for returning to the soil the excess calcium from ingested soil and food. The granules may also function in maintaining acid–base balance for animals in anaerobic environments.
4. In the Serpulidae, the calcareous granules contribute importantly to the formation of the tube in which the animal lives. The granules reach the exterior of the animal as a slurry that is shaped by the collar as it solidifies. By extending the open end of the tube in this way, the animal has in effect devised an exoskeleton that lengthens proportionately to the growth of the body.
5. The tube matrix contains acid mucopolysaccharide. The mineral is aragonite, low magnesian calcite, or high magnesian calcite.
6. There is uncertainty about the site of granule formation in serpulids. In some species, it appears that they form extracellularly in the lumen of the gland; in others, they may form in intracellular vesicles and then pass out of the cells to the lumen.

7. The opercular plate of serpulids consists of an outer cuticle and two calcified layers, all formed by a single layer of epithelial cells. This represents a second type of mineralization system, namely, an epithelial system that is also present in other phyla.

References

Bernhardt, A. M., Manyak, D. M., and Wilbur, K. M. (1985). *In vitro* recalcification of organic matrix of scallop shell and serpulid tubes. *J. Moll. Stud.* **51**, 284–289.

Bubel, A. (1983). A fine structural study of the calcareous opercular plate and associated cells in a polychaete annelid. *Tissue Cell* **15**, 457–476.

Chapron, C. (1971). Étude du role de la glande de Morren des lombriciens dans la régulation de l'eau. *J. Microscopie* **10**, 351–356.

Clark, A. M. (1957). The distribution of carbonic anhydrase in the earthworm and the snail. *Aust. J. Sci.* **19**, 205–207.

Crang, R. E., Holsen, R. C., and Hitt, J. H. (1968). Calcite production in mitochondria of earthwarm calciferous glands. *Bioscience* **18**, 299–301.

Darwin, C. R. (1881). "The Formation of Vegetable Mould through the Action of Worms, with Observations on Their Habits." Murray, London.

Day, J. H. (1967). "A Monograph on the Polychaeta of Southern Africa. Part 2. Sedentaria." Trustees of the British Museum of Natural History, London.

Dotterweich, H. (1933). Die Function tierischer Kalkablagerungen als Pufferreserve im Dienste der Reacktionensregulation. Die Kalkdrüsen des Regenwurms. *Pfluegers Arch.* **232**, 263–286.

Gabbay, K. H. (1958). An investigation of the calciferous glands in *Lumbricas terrestris. Biol. Rev. City Coll. New York* **21**, 16–19.

Hedley, R. H. (1956). The secetion of the calcareous and organic components of the tube of *Pomatoceros triqueter. Q. J. Microsc. Sci.* **97**, 411–420.

Hedley, R. H. (1958). Tube formation by *Pomatoceros triqueter* (Polychaeta). *J. Mar. Biol. Assoc. U.K.* **37**, 315–322. Cambridge University Press.

Jamieson, B. G. M. (1981). "The Ultrastructure of the Oligochaeta." Academic Press, New York.

Laverack, M. S. (1963). "The Physiology of Earthworms." Macmillan, New York.

Lee, K. E. (1985). "Earthworms: Their Ecology and Relationships with Soils and Land Use." Academic Press, New York.

Morgan, A. J. (1981). A morphological and electron microprobe study of the inorganic composition of the mineralized secretory products of the calciferous gland and chloragogenous tissue of the earthworm, *Lumbricus terrestris.* L. *Cell Tissue Res.* **220**, 829–844.

Muzii, E. O. (1965). Personal communication.

Nakahara, H., and Bevelander, G. (1969). An electron microscope and autoradiographic study of the calciferous glands of the earthworm *Lumbricus terrestris. Calcif. Tissue Res.* **4**, 193–201.

Neff, J. M. (1967). "Calcium Carbonate Tube Formation by Serpulid Polychaete Worms: Physiology and Ultrastructure," Ph.D. thesis. Duke University, Durham, North Carolina.

Neff, J. M. (1971). Ultrastructural studies of the secretion of calcium carbonate by the serpulid polychaete worm, *Pomatoceros caeruleus*. Z. Zellforsch. **120**, 160–186.

Nott, J. A., and Parkes, K. R. (1975). Calcium accumulation and secretion in the serpulid polychaete *Spirorbis spirorbis* L. at settlement. *J. Mar. Biol. Assoc. U.K.* **55**, 911–923.

Oglesby, L. C. (1978). Salt and water balance. In "Physiology of Annelids" (P. J. Mill, ed.), pp. 555–658. Academic Press, New York.

Piearce, T. G. (1972). The calcium relations of selected Lumbricidae. *J. Anim. Ecol.* **41**, 167–188.

Robertson, J. D. (1936). The function of the calciferous glands of earthworms. *J. Exp. Biol.* **13**, 279–291.

Satchell, J. E. (1967). Lumbricidae. In "Soil Biology" (A. Burges and F. Raw, eds., pp. 259–322. Academic Press, New York.

Stephenson, J. (1930). "The Oligochaeta." Clarendon Press, Oxford, England.

van Gansen-Semal, P. (1959). Structure des glandes calciques d'*Eisenia foetida* Sav. Bull. Biol. (Woods Hole, Mass.) **93**, 38–63.

Vovelle, J. (1956). Processus glandulaires impliqués dans la reconstitution du tube chez *Pomatoceros triqueter* (L.) Annelide Polychète (Serpulidae). *Bull. Lab. Marit. Dinard* **42**, 10–32.

Wiecek, C. S., and Messenger, A. S. (1972). Calcite contributions by earthworms to forest soils in Northern Illinois. *Soil Sci.* **36**, 478–480.

13

Crustacea—The Dynamics of Epithelial Movements

The animals commonly recognized as members of the subphylum Crustacea are the crabs, lobsters, shrimp, crayfish, and barnacles. However, the great majority of crustacea are extremely small organisms present in the oceans. Because of their size and incredible numbers, it has been estimated that the average size of crustacea is of the order of 1 mm. The environments inhabited by crustacea include shore waters, the open sea and abyssal depths, and fresh waters, while a few are terrestrial.

The body and the appendages of crustacea are covered by the carapace, a chitinous or chitinoid material that in most species is mineralized with calcium carbonate. The mineral deposition in some groups may be very extensive, as in the barnacles, which form rigid calcareous plates (Fig. 13.1) and in the ostracods, which are enclosed in a bivalve shell. As a result, the crustacea typically undergo a complex series of changes as the exoskeleton is resorbed and reformed during periods of intermittent growth.

Well over 250 years ago, the French biologist Réamur described the process whereby the exoskeleton of crustacea is shed and replaced by a new one. As the animal grows, the process is repeated in cyclic fashion, each new exoskeleton being slightly larger than the last. The cycle of molting is accompanied by a cycle of calcification in which the old one is decalcified prior to shedding and the new exoskeleton becomes mineralized soon after its formation. This remarkable cycle of mineralization is of particular interest in studies of biomineralization for four reasons.

Figure 13.1 Surface and cutaway view of a typical barnacle shell. Arrows indicate directions of growth: (1) downward growth at the basal margin, (2) growth at the margin of the radii, (3) growth at the margin of the alae, (4) growth at the margin of the operculum (Bourget, 1980).

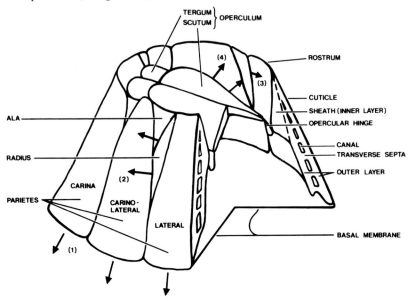

First, the heavily calcified exoskeleton of aquatic forms, such as marine crabs and lobsters or the freshwater crayfish and shrimps, provides a clear example of the membrane pumps involved in accumulating the ions of mineralized skeletons. Second, the occurrence of the molt cycle demonstrates the involvement of an epithelial layer that is capable of both the resorption and secretion of a complex mineralized structure within a short period. Third, a portion of the minerals that are resorbed from one skeleton are stored within the body in a number of ingenious systems and are reused a short time later in the mineralization of the new and enlarged exoskeleton. Finally, the whole process is under complex endocrine control.

I. The Mineralizing System

The exoskeleton of crustacea, commonly called the carapace or cuticle, consists of four layers: an outermost epicuticle and, in order beneath, the exocuticle, the endocuticle, and the membranous layer (Figs. 13.2, 13.3). The cuticle has three main components: protein, including glycoprotein, chitin, and calcium carbonate, usually calcite. In addition, mucopolysaccharides and lipids are present. The proteins are of two kinds, a water-insoluble fraction called sclerotin and a water-soluble fraction called arthropodin. α-type chitin is linked to the protein. The calcium content varies with species and with the area of the cuticle. In several species, the calcium carbonate amounts to 50 to 80%, or slightly more, of the cuticle dry weight. Before the cuticle is shed in molting, the calcium content is sharply reduced (Table 13.1).

The study of the fine structure of the cuticle has long been actively pursued and differences have been described for various species. We shall mention a few details of the four layers as seen in vertical section (Travis, 1963; Hegdahl *et al.*, 1977a–c; Roer and Dillaman, 1984).

A. Epicuticle

The epicuticle, the thinnest of the four layers, is made up of lipoprotein tanned by polyphenol oxidase. It consists of two layers, and has spines. The pore canals (see below) within the lower layer and the spines are mineralized internally. Spherulitic aggregates have also been described.

Figure 13.2 Diagram of cross section of crustacean carapace. Note branching of pore canals. (After Compère and Goffinet, 1987.)

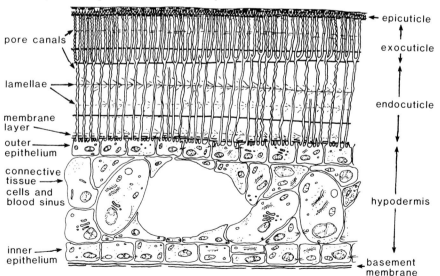

Figure 13.3 (a) Phase-contrast micrograph of horizontally sectioned decalcified exocuticle of *Cancer pagurus* in the intermolt condition. Dark polygonal cross-sections are prisms with long axes perpendicular to the face of the section. (From Hegdahl *et al.*, 1977b.) (b) Electron micrograph of decalcified preparation from the endocuticle of *Cancer pagurus* in the intermolt condition showing longitudinally sectioned fibrils. (From Hegdahl *et al.*, 1977a.) (c) Electron micrograph of exocuticle of the leg of *Astacus fluviatilis* decalcified in EGTA. The exocuticle is basically helicoidal with a system of reticulate rods and spaces running through it. (From Neville, 1975.)

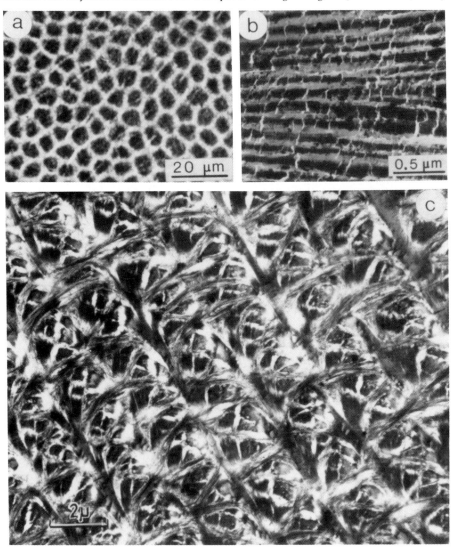

observed that the crystal plates showed mosaic patterns in polarized light, indicating that nucleation and growth were occurring at numerous sites.

Mineral deposition has been extensively studied in decapods. In the crab, mineral first appears around the pore canals and then throughout the chitin–protein fibrillar network (Yano, 1980). A variety of techniques including electron microscopy (Roer and Dillaman, 1984), secondary ion mass spectrometry, and X-radiography (Giraud-Guille and Quintana, 1982) indicate that the crystals are aligned with the organic fibers of cuticle. According to Hegdahl *et al.* (1977a,b) and Roer and Dillaman (1984), rod-shaped crystals are interspersed with the chitin and protein fibers in the cuticle of *Carcinus maenas*.

The relationship of magnesium and phosphate ions to the crystallinity of the mineral in the crustacean cuticle has been a subject of interest for many years (see Richards, 1951). It has been repeatedly implied that there is a large amorphous component. Certainly in the gastroliths of the crayfish *Orconectes virilis* the calcium carbonate appears to be entirely amorphous (Travis, 1963), while the calcium carbonate granules found between the midgut cells of the amphipod *Orchestia carcinana* are also without any crystallinity (Graf and Meyran, 1985). Both deposits are involved with fluxes of calcium during the molt process, and since the amorphous state is much more soluble than the crystalline state, more attention should be given to the relationship between mineral state and the rates of calcification and decalcification.

In many decapods, amorphous deposits of calcium phosphate are found at intracellular sites in the hepatopancreas (Simkiss, 1976).

III. Ion Transport

In the postmolt stage, most crustacea take up calcium from the water in which they live, and Greenaway (1985) has identified two types of pumps. In freshwater species, the uptake mechanism has a high affinity for calcium (K_m in the range of 0.1 to 0.3 mmol/liter) and saturates typically at a relatively low concentration of 2.0 mmol/liter. Marine species have a lower affinity (K_m of 10 mmol/liter) and saturate at levels closer to the high calcium concentrations of seawater (saturation 13 mmol/liter). These uptake mechanisms appear to be localized in the gills and are probably driven by a Ca-ATPase. Since they are not electrogenic, they probably involve antiport systems of the $Ca^{2+}/2H^+$ type that move calcium into the blood.

The immediate source of ions for mineralization of the cuticle is the hemolymph which supplies the epidermis. The total calcium concentration of this fluid is typically in the range of 8 to 15 mmol/liter, although it does vary during the molt cycle (Greenaway, 1985).

The mechanisms for the deposition of $CaCO_3$ in the cuticle, and for its demineralization before molting, reside in the outer epithelial layer of the epidermis beneath the membranous layer (Fig. 13.2). As we mentioned, these cells form numerous cytoplasmic extensions which penetrate the cuticle as far as the epicuticle. The number of cytoplasmic extensions in different species has been estimated to be between 150,000 and 4,000,000 per mm^2, and the holes that they occupy are referred to as pore canals (Roer and Dillaman, 1984). Because of their number and length, these extensions provide an enormous cell surface for transport activities (Roer, 1980). During the mineralization of the exoskeleton, they are thought to transport calcium outward across their membranes into the cuticle. The pore canals themselves are commonly mineralized, with presumably most of the deposits being external to the cell membrane. It is probable that the cell extensions also secrete organic material during cuticle formation.

Experiments on preparations of the isolated epidermis of *Carcinus maenas* have provided information on the transport properties of the epithelial cells (Roer, 1980). A Ca-ATPase pump apparently moves calcium outward through the apical cell membrane (Fig. 13.4) which also contains a Na^+/Ca^{2+} system capable of exchanging intracellular Ca^{2+} or Na^+ for extracellular Ca^{2+} or Na^+. These types of pumps and exchange systems are not specific for crustacean epithelia and are found in a variety of other cell types as well.

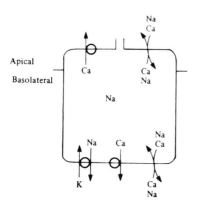

Figure 13.4 Proposed model of the net calcium uptake by postmolt *Carcinus maenas* integument under normal conditions in control media. (From Roer, 1980.)

Roer (1980) has proposed an ingenious hypothesis for the manner in which the outer layer of the epithelial cells of the epidermis accomplishes an outwardly directed calcium flux during mineralization of the cuticle and inwardly directed calcium resorption prior to molting. Consider first the mineralization of the cuticle. The epithelial cells are in the squamous form but have the enormous extensions of their apical surfaces into the pore canals. This provides a very large area of membrane penetrating the cuticle layers (Fig. 13.2). By contrast, the basolateral area of these cells will be quite small. Assuming that calcium is pumped outward from the cytosol at a uniform rate by all parts of the cell membrane, it will be evident that the net calcium flux from the cell extensions within the cuticle will be much greater than that in the reverse direction toward the hemolymph. This will favor mineralization of the cuticle. Then, during the period preceding a molt, when the epithelial cells separate from the old cuticle and the cell extensions are severed from the apical cell surfaces, there will be a drastic reduction in the size of this Ca-pumping membrane. The epithelial cells now change shape from squamous to columnar. Consequently, the area of their basolateral surfaces now exceeds that of the apical surfaces. Assuming once again equal pumping rates in all areas of the membrane, the net calcium flux will be predominantly toward the hemolymph and demineralization of the cuticle will accordingly be favored. Hormonal effects initiate both the cellular transformations from a squamous to a columnar epithelium and also the other changes associated with the molting process (Skinner, 1985).

IV. Mineralization

Before crabs molt, they deposit the two outer layers of their new cuticle, and these remain uncalcified until after molting when they become calcified within a few hours. Roer and Dail (cited in Roer and Dillaman, 1984) examined the capacity of the pre- and postmolt cuticle to calcify *in vitro* by stripping off the hypodermal tissue of pieces of both types of cuticle followed by fixing, decalcification, and placing in a calcifying solution. Mineral deposition occurred in the postmolt cuticle but not in the premolt cuticle. The results suggest that, following molting, the hypodermis secretes nucleating matrix that initiates crystal formation in the cuticle or that an inhibitor of mineralization is removed. After the molt, the newly deposited endocuticle is calcified along with the epicuticle and exocuticle, which were formed somewhat earlier. The deposition of $CaCO_3$ is apparently caused by ion movements from the cell extensions within the various layers, as we have already mentioned. The extremely close spacing of the extensions would make possible a relatively rapid attainment of the concentration required for mineralization. The calcite crystals and crystal aggregates that are deposited will be interspersed among the chitin–protein fibrils that are organized into lamellae. The pore canals and the spines of the epicuticle will also be mineralized. Since the organization of the organic material differs among the layers, it is not surprising that the form taken by the mineral is also different.

Details of the mineralization process are not clear beyond the fact that the microspaces among the cuticle meshwork apparently govern the form of the crystals that are deposited. On the basis of ruthenium red staining, Giraud-Guille (1984) has proposed that crystal nucleation may be brought about by anionic groups on glycoproteins of the fibril networks. In view of the enormous surface of the networks, nucleation might well be expected even in the absence of anionic groups.

If, as seems likely, the CO_3^{2-} of the $CaCO_3$ mineral is formed from HCO_3^- moving out from the cell extensions, then protons will be released into the cuticle meshworks. If these protons were to accumulate, further precipitation of $CaCO_3$ would be inhibited. In fact, it appears that the opposite occurs. By measuring the pH of the shell fluid with the indicator DMO, Cameron and Wood (1985) were able to show that it was 0.3 to 0.5 pH unit above that of the blood, i.e., it was alkaline. During the period of carapace mineralization, the crab *Callinectes sapidus* excretes protons and accumulates calcium at rates of about 12 meq. kg^{-1} h^{-1}. These are such high rates that it appeared unlikely that the crab could actually derive the carbonate ion from metabolic carbon dioxide. Subsequent quantification of these ion fluxes showed that three protons were excreted for every two calcium ions absorbed (Cameron, 1985). The overall reactions appeared to be

$$
\left.
\begin{aligned}
\text{(Metabolic) } CO_2 + H_2O &\rightarrow CO_3^{2-} + 2H^+ \\
\text{(Uptake) } \qquad HCO_3^- &\rightarrow CO_3^{2-} + H^+ \\
\text{(Uptake) } \qquad 2Ca^{2+} &\rightarrow 2Ca^{2+}
\end{aligned}
\right\} 2CaCO_3 + 3H^+
$$

The hydration and dehydration of carbon dioxide, which are so important in many acid–base reactions, are catalyzed by carbonic anhydrase, which is known to be present within the crustacean cuticle (Giraud, 1981). Thus the epidermis appears to be well adapted for rapid mineralization but the supply of calcium along the length of the cell extensions and the elimination of protons and CO_2 may well require some sort of circulation of cytosol into and out of these cellular extensions. This possibility has not been studied experimentally.

The innermost layer of the cuticle, the membranous layer, and the inner portion of the epicuticle, except for the pore canals, lack mineralization. This is perplexing in view of the presence in both layers of cell extensions that could supply calcium and organic material. The hypotheses usually put forth to explain an inhibition of mineralization are the presence of inhibitory material and the lack of nucleation sites. Information is lacking in the Crustacea for both these schemes.

V. Calcium Storage

During the premolt period, the extensive resorption of the cuticle liberates large amounts of calcium which are, either lost to the environment or stored within the body of the animal (Table 13.1). Both the percentage of calcium lost and the percentage stored vary greatly among species. The range of values for stored calcium is 4 to 75%. The storage mechanisms are particularly interesting in that they provide systems for the cyclic accumulation and release of calcium, which are synchronized with demineralization and mineralization of the cuticle, respectively.

The storage sites are gastroliths, midgut glands (hepatopancreas), midgut cecae, and hemolymph (Greenaway, 1985). The gastroliths are disk-shaped calcified structures which form in the anterior wall of the cardiac stomach between an epidermis and a cuticle (Fig. 13.5) (Travis, 1963). The mineral that is deposited here is poorly crystalline calcium carbonate. At the time of its deposition, which coincides with the demineralization of the cuticle, the epidermal cells increase in height, become branched, and provide the organic matrix of the gastrolith. According to Ueno (1980), the main function of the tissue associated with the gastroliths is to transport calcium from the blood into the mineralized deposit. He suggests that they do this by accumulating calcium in the mitochondria, which are then sloughed into the lumen of the stomach to provide calcium as they degenerate. It is more likely, however, that the rapid calcium flux through the gastrolith tissue results in an increased penetration of calcium into the cells and that the mineralized mitochondria are a result of this rather than the cause of gastrolith calcification. Greenaway (1985) notes that generally the calcium stored in gastroliths is a very small fraction of the calcium in the body and it can supply less than 10% of the amount required for the mineralization of the new cuticle. In many marine crabs; calcium is stored in the hepatopancreas as calcium phosphate granules, and the mitochondria have again been implicated in this process (Chen et al., 1974). The same con-

hemolymph has been considered a storage site for the calcification of the new cuticle. Greenaway (1985) has pointed out, however, that the increase is usually of the order of a few micromoles and is insignificant in providing calcium for cuticular mineralization. There is, however, at least one striking case in which the hemolymph *does* function as a significant calcium store. Toward the end of the premolt stage in the freshwater land crabs of the genus *Holthuisana*, the hemolymph first becomes cloudy and then creamy white (Sparkes and Greenaway, 1984). Its calcium level increases some 150 times above the intermolt value. The major part of this calcium exists as microspheres 0.25 μm in diameter suspended in the blood. They consist chiefly of $CaCO_3$ with traces of phosphate and magnesium and an organic content of 22%. Their site of formation is not known, but it has been calculated that they must be formed at a rate of 46 million per second per gram of body tissue.

These systems of calcium storage are under close physiological control, and this is clearly seen in the wood louse *Oniscus asellus*. In this isopod, the posterior part of the exoskeleton is always shed before the anterior. When this happens the new posterior region is calcified by calcium which is derived from the stores in the anterior part of the exoskeleton. The decalcified anterior region is subsequently shed (Steele, 1982). The significance of this phenomenon is, of course, that the wood louse is able to utilize the mineral being resorbed from one part of the exoskeleton to mineralize another region. It is obvious, however, that in order to do this it must be able to regulate, simultaneously, both decalcification and calcification.

The dynamic nature of the crustacean exoskeleton will be apparent from these examples of adaptations to molting. But perhaps the most remarkable example of the versatility of this mineralizing system is found among the barnacles, where the system of molting has been adapted to a life-style where a sessile animal lives within a permanent shell.

VI. Barnacles

There are three orders of barnacles or cirripedes: some are parasitic, some burrowing, and only the Thoraceca have an external skeleton of calcareous plates. The latter are the forms that we will therefore examine in some detail. The earliest stalked barnacles simply developed armored plates but in the more advanced forms, dating from the upper Cretaceous, the body becomes completely enclosed. This raises two problems. First, how does the animal grow? As Darwin (1854) observed, if it is sealed to a rock and enclosed within plates, how can it expand in size? Second, if it is a true crustacean what happens when the animal molts? "A cirripede cannot like a crab crawl into some crevice and remain protected till its shell becomes hardened" (Darwin, 1854).

A. The Mineralizing System

The shell of the balanomorph barnacle is composed of calcareous plates, eight in number, joined in the form of a truncated cone (Fig. 13.1). The opening at the top is covered by two movable opercular plates which, when open, permit feeding by extension of the appendages. The side plates, called parietes, rest on a basal disk firmly cemented to the substratum. The disk may or may not be calcified (Bourget, 1980). With the exception of two genera reported to be aragonitic (Lowenstam, 1964), the mineral of the plates is calcite as shown by X-ray diffraction (Bourget, 1977) and electron diffraction (Nousek, 1984), whereas the calcified basal disk has been reported to be aragonite (Lowenstam, 1964). The way that the animal grows within these plates was a mystery that was solved by Darwin (1854), who realized that the individual shell plates could grow at their edges although these are often intricately overlapped. Thus, the plates increase in area by mineral deposition at their sides and lower edges. The base also increases in diameter by deposition at its edge. The result of these additions is an ever-increasing interior volume available for growth of the animal. In older animals, however, shell growth and the volume within may be restricted by fusion of the plates and by the downward growth of the plates below the edge of the base, thus preventing its increase in diameter (Crisp and Bourget, 1985).

How then is this growth achieved and what happens when the barnacle molts? Darwin and others supposed that the barnacle shell was simply a calcified exoskeleton in which successive layers were secreted by the epidermis and retained within each other instead of being shed. However, if this were so, the shell would be laminated with sheets of cuticle representing each molt, and this does not appear to be the case (Costlow, 1956; Bourget, 1987). Instead, it appears that the epidermis has become specialized. Whereas in other crustacea the epidermal cells each secrete the various components of the cuticle in succession, in the barnacles the various regions of the epidermis have become specialized for different products (Fig. 13.7B).

In the center of the base plate of *Balanus balanoides* the epidermis

Figure 13.7 Growing edge at the basal margin of the barnacle shell plate. The cuticular covering does not show the different constituent layers. Each cuticular scale is produced during one molting cycle. Scales overlap one another on the outer surface of the shell plate. →, Direction of movement of the newly formed cuticular slip; ➡, region of formation of the cuticle (A-A¹); ⇨, region of CaCO₃ deposition (B → B') (Bourget, 1980).

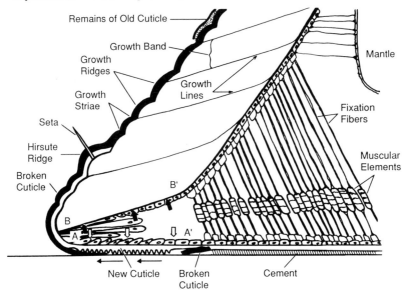

shows relatively little secretory activity, although some species do have a calcified base. The barnacle is, however, cemented to the substrate with a viscous cement secreted through ducts and provided with sufficient slippage to enable the animal to slide if pushed by adjacent individuals. The attachment of the barnacle is largely by fixation fibers and muscles (Fig. 13.7).

The epidermis lying outside the insertions of the fixation fibers retains the ability to secrete a chitinous exocuticle. The secretory activity is not continuous, however, but is synchronized with the animal's molting cycle. At this time a rather folded new cuticle is secreted at the basal margin of the shell. During the premolt stage the epidermal cells beneath the shell also secrete calcium carbonate (Fig. 13.7). The shell therefore increases in thickness at its base and moves outward, extending the circumference of the animal, breaking the old cuticle, and unfolding the new. In essence, therefore, there is a production of cuticle at the outermost region of the base of the shell which allows the barnacle to increase in diameter. Calcification at the base of the plates increases their size and there is a slight increase in calcium deposition over the whole of the epidermis underlying the shell (Crisp and Bourget, 1985).

The crystal structure of the shell that is found varies greatly among different genera. This was shown by Bourget (1977, 1987) in an examination of the fine structure of shells of 14 genera. In most shell plates the c-axis of the crystals is commonly orthogonal to the plane of the epithelium depositing the crystals. However, in some areas there are both large and small crystals, but without any uniform orientation.

Organic matrix is clearly evident in shell sections and may take various forms depending on the shell area and the genus (Bourget, 1977, 1987; Nousek, 1984; Crisp and Bourget, 1985). Single crystals may be enveloped by matrix; sheets of matrix may cover relatively large crystal structures and may or may not conform to the orientation of the crystals (Bourget, 1977); and interconnected ropelike fibers may form mats. It is difficult to relate these matrix components to homologous regions of a normal crustacean cuticle, but if the outermost layer corresponds to the epicuticle, then the calcified shell is probably endocuticle. There is certainly some evidence that it contains arthropodin and chitin components (Crisp and Bourget, 1985).

B. Mechanisms of Mineralization

Calcium presumably enters the shell aperture and passes into the mantle through its inner epithelium. Once inside, the calcium could enter the cells of the outer mantle epithelium and be pumped actively to the inner plate surfaces where crystals form and organic matrix is secreted.

The pattern of growth of shell plates can be seen in radial sections. *Chthamalus*, for example, has a so-called one-layer shell in which the growth is uniform as the plate extends downward (Fig. 13.8A). A plate of *Balanus*, on the other hand, is said to be two-layered because there are two major areas of growth (Fig. 13.8B). Other genera may have still different patterns. The regions and rates of past growth are reflected in the growth bands, wider bands indicating more rapid growth (Fig. 13.8B). The relative rates of growth in different areas of a plate can also be demonstrated qualitatively by exposure of the animal to ^{45}Ca and making autoradiograms of sections of the plate (Bourget and Crisp, 1975).

The form of shell plates as seen in cross section will depend on the rates of mantle growth and $CaCO_3$ deposition by the underlying epithelium. Consider the cross sections of the two types of plates in Fig. 13.8.

Figure 13.8   Distribution of the growth bands as seen in radial sections of a barnacle shell. (A) Bands continuous along the inner surface of the shell in *Chthamalus;* (B) bands discontinuous near the basal margin of the shell plate in *Balanus.* Note the absence of sheath in *Chthamalus* (Bourget, 1980).

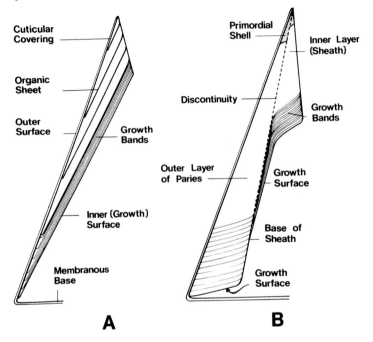

1. If mineral deposition occurs uniformly in all mantle regions, then the thickness at any point of the plate will be a linear function of the length of time it has been covered by the mantle. Expressed differently, if the rate of downward growth of the mantle is constant and the rate of mineral deposition remains constant, the thickness of the shell plate will show a linear decrease from the oldest portion to the bottom growing edge, as shown in Fig. 13.8A.
2. If mineral deposition is limited to the growing mantle edge and the edge advances at a constant rate, the thickness of the plate will be uniform. This is approximately the case in the lower portion of Fig. 13.8B.
3. If a limited part of the mantle edge deposits mineral but grows more slowly than the rest, then a secondary plate area will develop, as

shown in the upper right portion of Fig. 13.8B. If the secondary region of the mantle deposits at the same rate as other regions, then the secondary area will be thicker.

Condition 1 applies to molluscan shells in which shell thickness increases more or less uniformly with increase in size. Condition 2 is present in shells in which the thickness remains relatively constant with increase in size.

C. Growth Bands and Growth Rates

A discontinuity in the process of crystal formation in a barnacle shell will be evident in sections of the plates as a line. The spacing of such lines, called growth lines, serves as a measure of relative rate of plate growth and will reflect environmental conditions and experimental influences on mineral deposition. As shown in Fig. 13.8, the lines also indicate the position of the mantle at the time that crystal formation was interrupted. Growth increments bounded by two growth lines are referred to as growth bands (Bourget, 1980). *Balanus balanoides* present in the intertidal zone forms two bands per day. This is apparently due to cessation of growth on exposure to the air at low tide and its resumption when the tide comes in. The width of the growth bands corresponds with the period of immersion (Crisp and Richardson, 1975) and with continual submergence they are less distinct. An important factor limiting mineral deposition when a barnacle is out of the water appears to be the absence of calcium normally supplied by the seawater.

In addition to the growth bands in sections of the plates, growth bands can also be seen in some species on the external and internal plate surfaces and on the base. The concentric growth bands on the base of young specimens of the genus *Balanus* represent major zones of growth, and are especially interesting in that the crystal structure changes in a regular progression as each band is formed (Fig. 13.9) (Losada, 1974). Within each growth band, lines varying in number from 15 to 40 may be evident.

The growth bands of the base plates of *Balanus improvisus* differ from those resulting from tidal influences in that their frequency of formation is independent of tides, although they do show changes with other environmental conditions and with age (Losada, 1974). For example, the mean time required to form a band at 20°C was 1.8 days at 5 days of age and 2.9 days at 40–41 days of age. Areas of individual bands were used by Losada (1974) as a measure of growth rate over time. The areas in

Figure 13.9 Frontal view of portion of the basal disk of *Balanus improvisus* under polarized light. Note the concentric bands (b), the crystalline differences within a band, and the radial canals (r) (Losada, 1974).

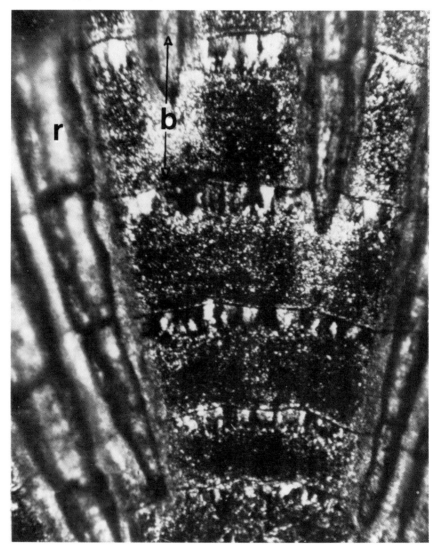

successive growth cycles followed a sigmoid curve: a slow rate during the first week, a more rapid increase from the second to the fourth week, and then a decrease in rate (Fig. 13.10). The area of cycles that occurred after those in Fig. 13.10 showed little or no increase. A plot of the *cumulative* area of the base versus age in days gives a similar sigmoid curve and is a pattern commonly observed in invertebrates.

D. Molting and the Control of Shell Growth

In most crustacea the calcification and resorption of the exoskeleton is synchronized with the molt cycle. The possibility that a similar phenomenon occurs in barnacles was examined by Bocquet-Vedrine (1965), who used *Elminius modestus* and observed that calcium is deposited at the growing edges of the shell plates during intermolt. Losada (1974) investigated the problem further by determining the correlation of molting frequency and growth band formation over three to five molt cycles in the basal plate of young *B. improvisus*. Sixty percent of the animals were

Figure 13.10 Area of individual basal bands corresponding to different successive growth cycles of *Balanus improvisus*. Each growth cycle is equivalent to a basal band numbered from the center outward. Mean area and its standard deviation are shown for each cycle (Losada, 1974).

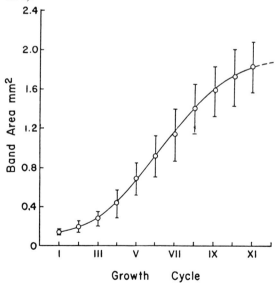

found to have a band/molt ratio of 1.00, and the remainder had ratios of 0.75 or 1.25. These findings are of considerable interest since they imply a link between the molt cycle of the carapace and mineral deposition by the mantle epithelium. It seems that, as with other crustacea, the hormonal control of molting extends to the regulation of mineral deposition by the mantle epithelium and perhaps to matrix secretion as well. There are, however, two aspects of the barnacle exoskeleton that appear unlike those of other crustacea. Thus, there is no evidence of pore acanals and no indication of any resorption of the barnacle shell prior to molt.

Bourget (1987), in discussing factors that influence barnacle growth, states that the production of epicuticular and cuticular material is correlated with molting but that shell deposition is related to tidal cycles and an endogenous rhythm. These latter influences do not of course, preclude other factors. In fact, it would appear probable that cuticular material formation and mineral deposition are interrelated through a biochemical pathway common to both. The findings now at hand indicate that further study of the relation of hormomal action to $CaCO_3$ deposition is needed in clarifying mechanisms of mineralization in this group of Crustacea.

VII. Summary

1. The rate of calcium metabolism in crustacea is much more active than in other invertebrate phyla because of the molting processes associated with growth (Greenaway, 1985). As a result, (a) cyclic demineralization of the old cuticle is followed by mineralization of the newly formed cuticle. (b) The filling and emptying of mineral storage sites are synchronized with decalcification and mineralization of the cuticle. (c) The shifts in direction of ion movement at the cuticle and at the storage sites are preceded by a transformation of the structure of epithelial cells responsible for ion movement. That is, cells transporting ions in one direction are different structurally from the same cells transporting ions in the opposite direction. Hormone action is undoubtedly responsible for this transformation in cell structure.

2. Mineralization of layers of cuticle far removed from the epithelium supplying the ions is made possible by extremely large numbers of closely spaced long cytoplasmic extensions. As a result, the distance for diffusion within the cuticle is short. The cell membrane sur-

rounding the extensions must actively transport calcium outward to reach a concentration exceeding the solubility product.

3. Sites in the cardiac stomach, midgut glands, midgut cecae, and the hemolymph temporarily store 4 to 75% of the calcium dissolved from the cuticle before molting.

4. The form of deposited mineral in crustacea appears to be largely governed by the spaces within the chitin–protein network of the cuticle. In the cuticle, the mineral is calcite but in the storage sites it is amorphous or poorly crystalline calcium carbonate or calcium phosphate.

5. The use of *in vitro* preparations of tissues, cuticle, and storage systems holds promise for future research on crustacean mineralization. Aspects that would appear to be profitably studied include ion transport (Roer, 1980), mineral deposition and decalcification, and the influence of hormones.

6. Barnacles show a number of differences from other crustacea in the way that the mantle epithelial system forms rigid plates of $CaCO_3$ enclosing the organism. Mineral deposition at the edges of the plates and at the edge of the base on which the plates rest increases the volume of the enclosed living space, thus permitting body growth.

7. The mantle epithelial system of barnacles is similar to that of brachiopods and molluscs in that it constructs a rigid exoskeleton that grows incrementally by the extracellular deposition of crystals in an organic matrix. The ultrastructure of the shell in the three groups of organisms is, however, quite distinct.

8. The microstructure of the barnacle shell varies with taxa. The crystals are commonly oriented in relation to the mineralizing epithelium, although in some shell areas the crystals have no uniform orientation.

9. The organic matrix takes several forms: envelopes of single oriented crystals, sheets, and mats of interconnected fibers. The chemical composition and influence of the matrix on crystal orientation and organization have not been studied.

10. The markings on the shell plates and base are evidence of a decrease in mineral deposition rates, a temporary interruption of crystal growth, or an increase in the relative amount of secreted organic material at these interruptions. The causes of these changes at the molecular level have yet to be explained but certain factors with which they are associated have been mentioned. One is molting and

is presumably hormonally influenced; another is age; a third is an endogenous rhythm, which is probably metabolic in nature. These factors deserve further attention, and recently attached barnacles in which growth cycles and crystal changes can be clearly observed in the light microscope would appear to be favorable experimental material for such studies.

References

Bocquet-Vedrine, J. (1965). Etude du tegument et de la mue chez le Cirrepede opercule *Elminius modestus* Darwin. *Arch. Zool. Exp. Gen.* **105**, 30–76.

Bourget, E. (1977). Shell structure in sessile barnacles. *Nat. Can.* **104**, 281–323.

Bourget, E. (1980). Barnacle shell growth and its relation to environmental factors. *In* "Skeletal Growth of Aquatic Organisms" (D. C. Rhoades and R. A. Lutz, eds.), pp. 469–491. Plenum, New York.

Bourget, E. (1987). Barnacle shells: Composition, structure and growth. *In* "Barnacle Biology" (A. J. Southward, ed.), pp. 267–285. A. A. Balkema, Rotterdam.

Bourget, E., and Crisp, E. J. (1975). Factors affecting deposition of the shell in *Balanus balanoides*. *J. Mar. Biol. Assoc. U.K.* **55**, 231–249.

Cameron, J. N. (1985). Postmoult calcification in the blue crab (*Callinectes sapidus*): Relationships between apparent net H^+ excretion calcium and bicarbonate. *J. Exp. Biol.* **119**, 275–285.

Cameron, J. N., and Wood, C. M. (1985). Apparent H^+ excretion and CO_2 dynamics accompanying carapace mineralization in the blue crab (*Callinectes sapidus*) following moulting. *J. Exp. Biol.* **114**, 181–196.

Chen, C. H., Greenwalt, J. W., and Lehninger, A. L. (1974). Biochemical and ultrastructural aspects of Ca^{2+} transport by mitochondria of the hepatopancreas of the blue crab *Callinectes sapidus*. *J. Cell Biol.* **61**, 310–315.

Compère, P., and Goffinet, G. (1987). Ultrastructural shape and three dimensional organization of the intracuticular canal systems in the mineralized cuticle of the green crab *Carcinus maenas*. *Tissue Cell* **19**, 839–858.

Costlow, J. D. (1956). Shell development in *Balanus improvisus* Darwin. *J. Morphol.* **99**, 359–398.

Crisp, D. J., and Bourget, E. (1985). Growth in barnacles. *Adv. Mar. Biol.* **22**, 119–244.

Crisp, D. J., and Richardson, G. A. (1975). Tidally produced internal bands in the shell of *Elminius modestus*. *Mar. Biol.* **33**, 155–160.

Darwin, C. (1854). "A Monograph on the Subclass Cirripedia, etc." Royal Society of London.

Giraud, M. M. (1981). Carbonic anhydrase activity in the integument of the crab *Carcinus maenas* during the intermolt cycle. *Comp. Biochem. Physiol. A* **69**, 381–387.

Giraud-Guille, M. M. (1984). Calcification initiation sites in the crab cuticle: The interprismatic septa. An ultrastructural cytochemical study. *Cell Tissue Res.* **236**, 413–420.

Giraud-Guille, M. M., and Quintana, C. (1982). Secondary ion microanalysis of the crab calcified cuticle: Distribution of mineral elements and interactions with the cholesteric organic matrix. *Biol. Cell* **44**, 57–68.

Graf, F. (1969). Le stockage de calcium avant la mue chez les Crustaces Amphipodes *Orchestia* (Talitride) et *Niphargus* (Gammaride hypoge). *Arch. Orig. Centre Doc., C.N.R.S.* **2690**, 1–216.

Graf, F., and Meyran, J. C. (1983). Premolt calcium secretion in the midgut posterior caeca of the crustacean *Orchestia:* Ultrastructure of the epithelium. *J. Morphol.* **177,** 1–23.

Graf, F., and Meyran, J. C. (1985). Postexurial calcium reabsorption in midgut posterior caeca of the crustacean *Orchestia carimarna:* Ultrastructural changes of the epithelium. *Cell Tissue Res.* **242,** 83–95.

Greenaway, P. G. (1985). Calcium balance and molting in the crustacea. *Biol. Rev.* **60,** 425–454. Cambridge University Press.

Hegdahl, T., Silness, J., and Gustavsen, F. (1977a). The structure and mineralization of the carapace of the crab (*Cancer pagurus* L.). I. The endocuticle. *Zool. Scr.* **6,** 89–99.

Hedahl, T., Silness, J., and Gustavsen, F. (1977b). The structure and mineralization of the carapace of the crab (*Cancer pagurus* L.). II. The exocuticle. *Zool. Scr.* **6,** 101–105.

Hegdahl, T., Silness, J., and Gustavsen, F. (1977c). The structure and mineralization of the carapace of the crab (*Cancer pagurus* L.). III. The epicuticle. *Zool. Scr.* **6,** 215–220.

Losada, F. J. (1974). "Studies on Growth and Shell Deposition in Barnacles of the Genus *Balanus,*" Ph.D. thesis. Duke University, Durham, North Carolina.

Lowenstam, H. A. (1964). Coexisting calcites and aragonites from skeletal carbonates of marine organisms and their strontium and magnesium contents. *In* "Recent Researches in the Fields of Hydrosphere, Atmosphere and Nuclear Geochemistry" (Y. Myake and T. Koyama, eds.), pp. 373–404. Maruzen Co., Tokyo.

Meyran, J.-C. Francois, J., and Graf, F. (1988). Analysis of the protein constituents of the calcareous deposits in the posterior caecae of the crustacean *Orchestia cavimana. Comp. Biochem. Physiol. B* **89,** 213–219.

Neville, A. C. (1975). "Biology of the Arthropod Cuticle." Springer-Verlag, New York.

Nousek, N. A. (1984). Shell formation and calcium transport in the barnacle *Chthamalus fragilis. Tissue Cell* **16,** 433–442.

Rhoads, D. C., and Lutz, R. A., eds. (1980). "Skeletal Growth of Aquatic Organisms." Plenum, New York.

Richards, A. G. (1951). "The Integument of Arthropods," pp. 1–411. Univ. of Minnesota Press, Minneapolis, Minnesota.

Roer, R. D. (1980). Mechanisms of resorption and deposition of calcium in the carapace of the crab *Carcinus maenas. J. Exp. Biol.* **88,** 205–218.

Roer, R., and Dillaman, R. (1984). The structure and calcification of the crustacean cuticle. *Am. Zool.* **24,** 893–909.

Simkiss, K. (1976). Intracellular and extracellular routes in biomineralization. *Symp. Soc. Exp. Biol.* **30,** 423–444.

Skinner, D. M. (1985) Molting and regeneration. *In* "Biology of Crustacea" (D. E. Bliss and L. H. Mantel, eds.), Vol. 9, pp. 43–146. Academic Press, New York.

Sparkes, S., and Greenaway, P. (1984). The haemolymph as a storage site for cuticular ions during premolt in the freshwater/land crab *Holthuisana transversa. J. Exp. Biol.* **113,** 43–45.

Steele, C. G. H. (1982). Stages of the intermolt cycle in the terrestrial isopod *Oniscus asellus* and their relation to biphasic cuticle secretion. *Can. J. Zool.* **60,** 429–437.

Travis, D. F. (1963). Structural features of mineralization from tissue to macromolecular levels of organization in the decapod Crustacea. *Ann. N.Y. Acad. Sci.* **109,** 177–245.

Ueno, M. (1980). Calcium transport in crayfish gastrolith disc; morphology of gastrolith disc and ultrastructural demonstration of calcium. *J. Exp. Zool.* **213,** 161–171.

Yano, I. (1980). Calcification of crab exoskeleton. *In* "Mechanisms of Biomineralization in Animals and Plants" (M. Omori and N. Watabe, ed.), 187–196. Tokai Univ. Press, Tokyo, Japan.

14

Molluscs—Epithelial Control of Matrix and Minerals

The Mollusca is a remarkably diverse phylum exhibiting adaptive radiation into a great variety of habitats. They inhabit the very deep and shallow seas, intertidal zones, and fresh and brackish waters; on land, they are found in the temperate zones and the tropics. The phylum comprises seven classes totalling some 110,000 species. Our discussion will be limited to representatives of the three major classes: gastropods, bivalves, and cephalopods.

The molluscs are mineralizers of great versatility, both in the variety of the mineralized structures that they form and in the types of minerals deposited. At least 26 structures other than shell are mineralized (Table 14.1), and they deposit 20 minerals (Table 14.2). The majority of these structures are formed extracellularly, although there are also intracellular granules. It is possible that some of the extracellular phosphate granules listed in Table 14.2 have an intracellular origin. A significant aspect of the great variety of mineralized structures listed in Table 14.1 is that each extracellular structure regardless of its complexity is formed directly by a single layer of epithelial cells. These cells must therefore be involved in both the movement of mineral ions to the site of deposition and in the secretion of organic matter that will become the matrix of the deposit.

The epithelial layer of the mantle is ideal for experimental studies of ion transport and matrix secretion as it is easily separated from the mineral it deposits. Without any permanent connection to the area of

Table 14.1 Molluscan Mineralized Structures[a]

Shell	Byssal complex
Epiphragms	Eye lens
Hinge ligaments	Spicules
Radula, teeth	Statoliths
Resilium	Darts
Calcified gill bars	Gizzard plates, beaks
Eggshells	Calcified arteries
Egg cases	Renal deposits
Spermatophore walls	Intracellular granules
Burrow linings	Extracellular phosphate granules of gills
Penial hooks, plates, stylets, rods, pallets	and inner shell surface

[a] According to A. S. Tompa (personal communication).

deposition, the mantle secretes a shell in which the minerals are deposited with remarkable precision and regularity. It is, moreover, the variety and complexity of the shells that are formed that make this system an exciting one for study.

I. The Mineralizing System

The system of shell formation can be illustrated in two ways, one showing an anatomical section through the mantle and shell of a bivalve (Fig. 14.1), and the other a diagram of the same structures represented as a series of compartments through which ions move (Fig. 14.2). Figure 14.1 shows a tanned protein layer called the periostracum (P) covering a shell with two types of crystal structures (PR, NC). The mantle or pallium is

Table 14.2 Molluscan Biominerals[a]

Mineral types	Numbers occurring
Carbonates	5
Phosphates	7
Halides	2
Iron oxides	4
Silica	1
Oxalate	1
	20

[a] After Lowenstam and Weiner (1983).

Figure 14.1 Radial section of the mantle edge of a bivalve to show the relationship between the shell and mantle. (Not drawn to scale.) EPS, extrapallial space; IE, inner epithelium; IF, inner fold; LPM, longitudinal pallial muscle; MC, mucous cell; MF, middle fold; NC, nacreous shell layer; OE, outer epithelium; OF, outer fold; P, periostracum; PG, periostracal groove; PL, pallial line; PM, pallial muscle; PN, pallial nerve; PR, prismatic shell layer.

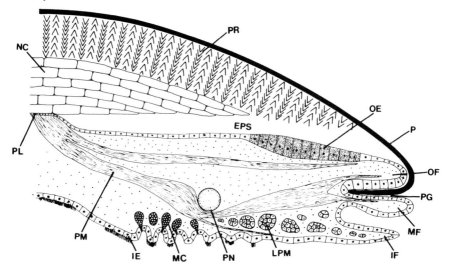

separated from the inner shell surface by a space (EPS) containing the extrapallial fluid. The edge and outer surface of the mantle are covered by an epithelial layer of various cell types responsible for formimg the periostracum and the individual shell layers. The inner mantle epithelium is in contact with the outer medium and may be an area for ion intake in aquatic species.

Figure 14.2 shows this system composed of six compartments. Compartment (1) is the outer medium (left) separated by the body epithelium (2) from the blood or hemolymph and tissues and (3) bounded by the mantle epithelium (4) facing the extrapallial fluid (5) and the shell (6). Two important points need emphasis. First, the ion movements may be bidirectional, both inward from the external medium and outward from the shell. Since the shell can provide ions to various parts of the organism, it is quite properly called a compartment. The second point is that bicarbonate ions may have a double origin, from the external medium and from the carbon dioxide of metabolism. With these concepts of serial compartments as background, it is possible to discuss ion move-

Figure 14.2 The molluscan mineralizing system. Note: (1) movement of Ca and HCO_3 from the medium toward the shell as occurs during shell deposition; (2) source of HCO_3 from the external medium and tissue metabolism; (3) movement of Ca and HCO_3 outward as occurs from shell dissolution; (4) secretion by the mantle of organic material of shell; and (5) ion exchange at the body epithelium (Wilbur and Saleuddin, 1983). (After Greenaway 1971.)

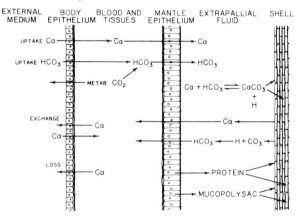

ments in greater detail (see Simkiss, 1976, 1984; Wilbur, 1984, 1985; Wilbur and Saleuddin, 1983).

II. Mechanisms of Mineralization

A. Ion Movements

Calcium in passing from the medium and through the epithelial cells of the outer body surfaces will either diffuse down an electrochemical gradient into the cytosol and then be pumped into the hemolymph against an electrochemical gradient or will pass through intercellular spaces. Since the calcium ion concentration of the hemolymph is much greater than that of most freshwater media, calcium ions will move up chemical gradients during its passage across the outer body epithelium and into the hemolymph. Once within the hemolymph, the calcium will be carried throughout the body and to the mantle, where it could pass across the outer epithelial layer to the inner shell surface, the site of mineral deposition. We may now consider ion movement across the mantle in more detail.

B. Calcium Movement Across the Mantle

The rate of calcium movement across the mantle epithelium of the oyster *Crassostrea virginica* was measured with radioisotopes by Jodrey (1953) in a preparation consisting of the isolated mantle attached to its shell. After the mantle had reached equilibrium, the rate of outward movement of ^{45}Ca was followed and found to agree closely with the rate of calcium deposition on the shell. It is interesting to note that the time taken for ^{45}Ca to penetrate the mantle compartments and reach equilibrium was about 2 hr for the mantle–shell preparation, with similar times for the mantle of the intact scallop *Argopecten* (Wheeler *et al.*, 1975) and the mantle of the intact shipworm *Bankia* (Manyak, 1982); however, a longer time was required for equilibrium in the clam *Mercenaria* (Dillaman and Ford, 1982). In another method for measuring ion movement, an isolated mantle preparation was arranged as a membrane between two solutions. The electrical potential across such preparations provides an indication of the ion movements. With the freshwater mussel *Anodonta* and the marine bivalve *Anomalocardia* the mantle resting potentials were 57 mV and 60 mV, respectively, shell side positive (Istin and Kirschner, 1968). With a probable free Ca^{2+} concentration of 10^{-7} to 10^{-8} M in the cytosol (Dipolo *et al.*, 1976; Borle and Snowdowne, 1982; Clarkson *et al.*, 1988) and a high Ca^{2+} concentration in the extrapallial fluid, the electrochemical gradient would be strongly against movement from the cells into the extrapallial fluid. That movement of calcium ions from the cells into the extrapallial fluid does occur suggests that they enter the extrapallial fluid by active transport; but this conclusion should be treated with care. Studies on the isolated mantle of the terrestrial snail *Helix* indicate that electrogenesis is probably due to active transport of chloride ions from the shell side to the hemolymph and that calcium and bicarbonate are not the primary ions that were being moved (Enyikwola and Burton, 1983). The movement of Ca^{2+} outward across the central mantle area of *Anodonta* has also been thought to occur by diffusion (Coimbra, 1988). These *in vitro* studies with their undoubted disturbance of normal transport conditions leave much still to be learned about calcium movement across the mantle of the intact animal.

It will be evident that active transport will be required to provide the supersaturation of ions necessary for crystal formation if calcium takes an intracellular route, as seems likely. In fact, an outward active transport of calcium by the cell membrane is a general characteristic of cells, and epithelia other than those of molluscs might well produce supersaturation in an enclosed microspace. Intercellular movement of calcium ions from the supersaturated hemolymph to the extrapallial fluid is possible in the central mantle area (Neff, 1972; Crenshaw, 1980), but the significance of this in the absence of concomitant active transport has yet to be established.

The energy changes necessary to move calcium across the mantle and induce crystallization on the inner shell surface are represented in Fig. 14.3. Calcium, on entering the epithelial cells from the hemolymph (H), diffuses down an electrochemical gradient (1). In order that Ca^{2+} can pass into the extrapallial fluid (EPF), energy, presumably supplied by ATP, must overcome the adverse electrochemical gradient (2). Crystal formation (S) then requires a further energy input in order to raise the ion activities above those of the solubility product (3). If, however, one assumes that the hemolymph and the extrapallial fluid have a similar free calcium ion concentration, then intercellular calcium movement to the extrapallial fluid would require no energy beyond that of diffusion and perhaps bulk flow (4).

C. Bicarbonate Movement

The calcium carbonate present in molluscan shell is derived from calcium and bicarbonate ions in the extrapallial fluid according to the reaction

$$Ca^{2+} + HCO_3 \rightarrow CaCO_3 + H^+ \tag{14.1}$$

The origin of the bicarbonate ions, i.e., whether from the medium, from metabolism, or both, can be determined experimentally by placing mol-

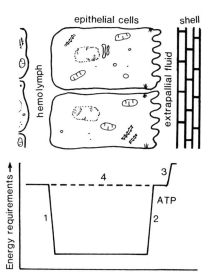

Figure 14.3 Schematic representation of the energy changes involved in the movement of calcium from the hemolymph into an epithelial mantle cell (1) and out into the extrapallial fluid (2) to crystallize onto the shell (3). The movement of calcium through the intercellular spaces avoids these energy barriers (4). (Modified after Simkiss, 1976.)

luscs in seawater containing $^{45}Ca^{2+}$ and $H^{14}CO_3^-$ and analyzing the $^{45}Ca/^{14}C$ ratio in the newly deposited shell (Wheeler *et al.*, 1975; Dillaman and Ford, 1982). If the ratio deposited in the shell is greater than in the seawater, then nonradioactive bicarbonate from metabolism has contributed to the shell mineral. In this manner it was demonstrated that the CO_3^{2-} of shell is supplied both from the HCO_3^- of the medium and from metabolism. In an examination of ion movement in the mantle of the oyster *Crassostrea virginica*, Wheeler (1975) found evidence for active transport of the bicarbonate ion. Further experimental studies of mantle transport of bicarbonate in parallel with calcium ions are clearly needed in order to determine both the mechanism and possible linkage between the movement of these two ions.

The calcium and bicarbonate ions that pass across the mantle epithelium are transported into the solution that lies between the mantle and the shell; and it is this extrapallial fluid that is the medium in which deposition of $CaCO_3$ crystals occurs.

D. Extrapallial Fluid

The inorganic ions of the fluid in the extrapallial space (Fig. 14.2) include not only Ca^{2+} and HCO_3^- but also Na^+, K^+, Mg^{2+}, Cl^-, and SO_4^{2-} (Crenshaw, 1972; Wilbur and Saleuddin, 1983). The calcium and carbonate ion concentrations are raised to levels exceeding the solubility product at the time of crystal formation. The pH of the extrapallial fluid of bivalves and gastropods from marine and freshwater environments is usually in the range of 7.4–8.3 (Wilbur, 1972), but this may change at the time of shell formation. The extrapallial fluid near the shell edge, where crystal formation is most rapid, occurs as a thin layer and this has made analyses very difficult. It is also possible that the transfer of ions and organic material from the mantle epithelium to the shell in this region is so intimate that it occurs virtually by direct contact. Unfortunately, the only region where the volume of extrapallial fluid may amount to as much as a few milliliters is in the central shell region of large bivalves. Because of its volume, this is the region from which fluid is normally taken for analysis, but it is quite likely that the composition at the shell edge differs considerably from this. For these reasons, it is necessary to speculate on what actually happens in the extrapallial fluid at the time of calcification.

It can be seen from Eq. (14.1) that, for each molecule of calcium carbonate that is formed, a proton is released. In order for the formation of mineral to continue, these protons must be removed. Wheeler (1975) has proposed that the protons react with bicarbonate ions and that carbonic anhydrase catalyzes the reaction. The carbon dioxide produced would then diffuse from the extrapallial fluid into the mantle epithelial cells. Other possibilities for removing protons remain to be examined, including an H^+-ATPase and an H^+/Ca^{2+} antiport. The active transport of monovalent ions outward from the extrapallial fluid into the mantle cells of *Helix* (Enyikwola and Burton, 1983) could also drive bicarbonate ions into the extrapallial fluid and so neutralize these protons. According to Campbell and Speeg (1969), ammonia may be synthesized at sites of calcification and these molecules would also remove protons by forming NH_4^+. Since, however, the distribution of total ammonia is pH dependent, it is difficult to distinguish cause and effect in this argument (see Wilbur and Saleuddin, 1983).

Under anaerobic conditions, succinic and other metabolic acids are formed in molluscs, lowering the pH of the extrapallial fluid (Crenshaw and Neff, 1969) and dissolving intracellular deposits of calcium carbonate and the inner shell surface (Wilkes and Crenshaw, 1979). The bicarbonate resulting from this dissolution provides buffering against these acids. This is clearly important in intertidal and other molluscs that are exposed to anaerobic situations.

III. Shell Matrix

The microstructure of molluscan shell is the result of an intimate linkage of crystals and organic matrix, both of which are derived from the extrapallial fluid. The matrix surrounds and separates individual crystals and is responsible for the layered crystal structuring of the shell. In addition to the structural coherence and the limitation of fractures that the matrix provides by its presence between crystals and crystal layers, four influences relating to shell growth have been attributed to it. These are crystal nucleation, crystal orientation, crystal type, and crystal size. Each of these effects is supported by some experimental evidence, yet none can be said to be firmly established.

The numerous studies of shell matrix comprise a sizeable literature, and a summary is scarcely warranted for our purposes. Rather, we shall mention very briefly a few properties relating to the presumed effects of matrix on crystals just mentioned and discussed in the sections that follow. More extensive accounts of matrix composition and functions will be found in Wilbur and Simkiss (1968), Krampitz and Witt (1969) Mann (1983), and Wilbur and Saleuddin (1983).

The matrix has its immediate origin in the secretion of the outer mantle epithelium and after release it undergoes physical and chemical changes during crystal deposition. It has long been known that its composition also differs between major shell layers, indicating that mantle areas responsible for forming these layers are specific in the material they secrete. The differences have been shown by histochemical methods (Ravindranath and Rajeswari Ravindranath, 1974) and by protein analyses of the individual layers.

The matrix of molluscan shell, as the matrix of many biomineral structures, can be separated in two major fractions, a soluble fraction and an insoluble fraction. Both fractions have been separated further and the subfractions analyzed. The principal compounds of the soluble fraction are glycoproteins, polypeptides, and polysaccharides. A general finding in the soluble matrix of shells of different species is a high content of aspartic and/or glutamic residues. These are considered important in providing carboxyl groups which bind calcium and may initiate crystal nucleation. Ester sulfate groups of glycoproteins are thought to act similarly. The insoluble matrix, in contrast to the soluble matrix, contains DOPA and has higher concentrations of phenylalanine and tyrosine and a greater total number of nonpolar residues (Meenakshi *et al.*, 1971; Ravindranath and Rajeswari Ravindranath, 1974; Wheeler *et al.*, 1988). Phenoloxidase has been reported to be present in shell (Samata *et al.*, 1980; Gordon and Carriker, 1980) and is presumably involved in the crosslinking of molecules in the formation of the insoluble matrix.

An examination of individual layers of nacreous shell by electron microscopy indicates that the growing crystals develop within soluble matrix since this matrix is present within the crystals (Watabe; 1965). Then, on attaining a particular crystal thickness, a layer of insoluble matrix layer is apparently added and then sclerotized *in situ*. This sequence of soluble followed by insoluble matrix, each with its distinctive composition, indicates that the two fractions of matrix are secreted alternately as each crystal layer is deposited (Crenshaw and Ristedt, 1976). The degree of precision of this control of secretion will determine the uniformity in layer thickness.

IV. Crystal Morphology

Molluscan shells exhibit a surprisingly wide and interesting array of crystal habits and patterns (Taylor and Kennedy, 1969; Taylor, 1973; Watabe, 1981, 1988; Carter and Clark, 1985). The more common crystal patterns have been classified somewhat problematically into the following five or six major types (for details, see Watabe, 1984, and Carter and Clark, 1985).

A. Prismatic Structures

These consist of parallel or fan-shaped elongate prisms enclosed in an organic sheath. (Fig. 14.4). Simple prismatic shell layers show that the elongate prisms are composed of crystalline disks or elongate blocks separated by organic matrix. Variations on simple prismatic structure have been classified as fibrous, composite, and spherulitic prismatic.

B. Spherulitic Structures

These are spherical structures of elongate subunits (crystallites) radiating from a discrete center (Fig. 14.5). Spherulites may develop into the prisms of prismatic shell and are often formed during shell repair.

C. Laminar Structures

Tablets, blades, laths, or rods in layered horizontal parallel arrangements are included in this category. Nacreous shell is the familiar example (Fig. 14.6). In most bivalves, the shell is built of tablets or blocks which, during development, form one or a few layers at a time (Fig. 14.7). The gastropods and cephalopods, on the other hand, form stacks of vertically aligned crystals in horizontal register (Fig. 14.8). As the crystals grow laterally and come into juxtaposition, layers are formed.

D. Crossed Lamellar Structures

These are tablet-shaped crystals, called first-order lamellae, composed of parallel rodlike second-order lamellae. Adjacent first-order lamellae are inclined at different angles but alternate lamellae are inclined at the same angle (Fig. 14.9). This structural type is termed simple crossed lamellar shell. Carter and Clark (1985) list 11 variations on this pattern.

E. Homogeneous and Granular Structures

Taylor and Kennedy (1969) designate *homogeneous* as a term of convenience for any fine-grained structure. The grains may be irregular or rounded and are usually less than 5 μm across. Aggregates of larger size are termed *granular*. These types of structures lack any well-defined

Figure 14.4 (a) Aragonite simple prisms of *Anodonta woodiana*. Note the interprismatic rigid wall enveloping the prisms. The preparation was etched briefly with EDTA after polishing (Suzuki and Uozumi, 1981). (b) Calcitic simple prisms of *Inoceramus grandis* from the Cretaceous. Radial fracture. (Courtesy of J. G. Carter.)

Figure 14.5 Developing spherulites of *Nautilus macromphalus* after 15 days regeneration of the shell (Meenakshi *et al.*, 1974).

Figure 14.6 Nacreous shell fracture surface of *Nautilus macromphalus*. The matrix between layers of crystals is not evident. (Courtesy of J. G. Carter.)

Figure 14.7 Deposition of crystals in step arrangement in growing nacreous shell of *Pinctada radiata*. The shell edge is upper left. Inset: Screw dislocation in newly deposited crystal of nacreous shell of same species (Wise, 1970.)

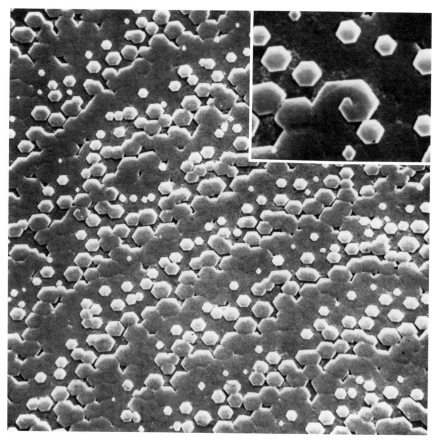

arrangement. However, in any single section of shell, their optical extinction in polarized light may be uniform (Carter and Clark, 1985).

Examples of taxa exhibiting various types of shell structures are given in Watabe (1984) and Carter and Clark (1985).

V. Crystal Patterns of Shell

In many species, the types of crystals that are deposited at any one time will be different in different parts of the inner shell layer. For example,

Figure 14.8 Developing nacreous shell of the gastropod *Turbo castanea* showing formation of stacks of crystals. Crystal size is a function of age, the topmost crystals being the most recently deposited (Wise, 1970).

crystals nearest the shell edge may be of an elongate crystal type with their long axis approximately perpendicular to the inner shell surface (i.e., prismatic shell) (Fig. 14.4), whereas in more central regions of the shell tabular crystals (i.e., nacreous shell) (Fig. 14.6) are being deposited at the same time. Clearly, the form of the crystals must not only be controlled by the activities of the epithelial cells that deposit them but must also be localized. Evidence for localized organic secretions comes from differences in the composition of the organic matrix from shell layers differing in crystal form in a single valve (Wilbur and Simkiss,

Figure 14.9 Cross lamellar shell of *Conus virgo*. Fracture surface showing lamellae. Note that adjacent lamellae are at roughly 90° angles. 22 mm = 50 μm. (Courtesy of K. Brear).

1968; Ravindranath and Rajeswari-Ravindranath, 1974; Weiner, 1983). From these compositional differences, there are strong indications that the physical and/or chemical nature of the matrix controls the crystal form. Since the crystals grow within the matrix medium, it follows that the form and direction of growth of these crystals will be governed, at least in part, by the molecular organization of the matrix (Chap. 2, III). Accordingly, the orientation of matrix molecules in prismatic shell can be expected to be different from the orientation of matrix molecules surrounding crystals of nacreous shell.

The control of crystal form and pattern is also emphasized by the marked differences among prismatic, lamellar, and crossed lamellar shell and variations within the individual types in different species (Carter and Clark, 1985). The structure of crossed lamellar shell provides an impressive example of control. In adjacent first-order lamels, the second-order lamels are oriented at an angle of 90° to 130° or more

(Bandel, 1979) (Fig. 14.9). During rapid growth, the snail *Ampullarius* may deposit on the average a first-order lamel every 40 min (Zischke *et al.*, 1970). If the organic material between the lamellae is the controlling factor in the crystal orientation, then the organic molecules secreted by the epithelial cells must be deposited at controlled and alternating angles at the same frequency.

Evidence for a causal relationship between crystal orientation and organic matrix will come at the molecular–crystal level. The relationship has been examined by means of X-ray and electron diffraction in lamellar shell structure, which has a relatively simple pattern of layers of single crystals in parallel arrangement with matrix between the crystal layers (Weiner and Traub, 1980, 1981; Weiner, Talmon, and Traub (S. Weiner, personal communication). A good alignment between insoluble matrix components and crystallographic axes was demonstrated in several species. For example, the nacreous layer of the septum of *Nautilus repertus* showed a well-oriented β-chitin pattern with the fiber axis parallel to the matrix between crystal layers and a β-pleated sheet with side chains perpendicular to the matrix layer. Electron diffraction of the shells of the oyster *Pinctada* and the gastropod *Tectus* has shown similar interrelations, with the *a* crystal axis aligned with the chitin and the *b* crystal axis aligned with the polypeptide chains. The orientation of the crystals is thought by Weiner and Traub (1980, 1981) to be brought about by soluble protein attached to an insoluble protein sheet. The initial binding of calcium ions by a $(Asp-Y)_n$-type sequence (in which Asp is aspartic acid with COO^- groups which bind Ca^{2+}, and Y is another single amino acid) is thought to provide sites of crystal nucleation. The bound Ca–Ca spacing is presumed to match some multiple of the spacing of the crystal lattice which is thereby induced. Following calcium binding, nuclei of calcium carbonate would be formed in association with these matrix molecules. Because of this relationship, the matrix molecules would become a template for epitaxial-type growth of the developing crystals. In this manner, crystal orientation could be controlled and crystal form influenced, although the degree of precision that is required by these processes is not clear. Mann (1983) points out that orientation may still occur under various conditions with a high degree of mismatch between substrate molecules and crystal lattice. Thus, crystals may also become oriented without specific epitaxial growth if they develop within a three-dimensional framework of oriented molecules. This type of orientation control should be considered further since molluscan shell formation provides an example where crystals develop *within* a matrix medium.

VI. Formation of Shell Layers

An examination of nacreous and crossed lamellar shell clearly shows that the crystals are deposited in layers (Figs. 14.6 and 14.9). Prismatic shell, on the other hand, is formed of elongate crystals at approximately right angles to the inner shell surface. We now consider the formation of nacreous and prismatic layers.

A. Nacreous Layer

The substrata of the first layers deposited in the adult shell are the last deposited inner layers of the larval shell and the periostracum formed by the mantle at the shell edge. Both substrata will probably be covered by organic matrix secreted by the overlying mantle epithelium. The transported calcium and bicarbonate ions, on reaching concentrations exceeding the solubility product in the extrapallial fluid, will form crystal nuclei scattered over the matrix surface. In the formation of a single nacreous layer in bivalves, the crystals that develop from the nuclei will come into contact with neighboring crystals as they grow and will fuse to form larger crystals (Fig. 14.7). Finally, the growing crystals will occupy the available space and a layer one crystal in thickness will be complete. Figure 14.7 indicates the manner in which these individual layers increase in area as the mantle grows.

The method of formation of nacreous layers is somewhat different in gastropods and bivalves, although both begin with a random distribution of small crystals. In gastropods, crystals of a second layer are nucleated at the centre of the upper surface of underlying crystals before those crystals have come into contiguity with neighboring crystals. A nucleus then forms at the centre of the surface of each superposed crystal and so on until pyramidal stacks of crystals are formed (Fig. 14.8). The nucleation of a crystal in the centre of the surface indicates that the site of formation is determined by organic material covering that surface. In some preparations, organic sheets can be seen bridging the gaps between crystal stacks. It is tempting to suggest that the presence of a crystal beneath the sheet could alter its molecular structure in such a way that a nucleation site would result. The secretion of the sheet covering many stacks of crystals could also have the effect of terminating growth in thickness of the crystals. In this manner, the sheet could keep all stacks in any given shell area in register. Thus, as the crystals grow laterally, they would make contact with corresponding crystals in neighboring stacks and individual layers would be formed. In bivalves, central surface nucleation on crystal surfaces does not occur and several layers may form simultaneously (Fig. 14.7).

Individual layers of nacreous shell are relatively uniform in thickness (Fig. 14.6), indicating that, at a given stage, growth of the lattice is interrupted. A new superposed layer is then initiated as just described. This sequence of layer initiation and termination at a particular thickness could result from more than one cause. The usual explanation suggests that the growing crystal lattice is inhibited by the secretion of organic matrix onto it. The evidence for this is based on two types of observations: (1) the presence of interlamellar matrix seen in vertical sections of shell, and (2) the inhibition of $CaCO_3$ formation *in vitro* by soluble organic matrix (Wheeler *et al.*, 1981). Extrapallial fluid, which bathes the inner shell surface and contains matrix constituents, also inhibits $CaCO_3$ formation *in vitro* (Wilbur and Bernhardt, 1984). Here, then, is an anomalous situation in which the medium in which crystals normally grow has the ability to inhibit crystal growth. One way out of this paradox was provided by Crenshaw and Ristedt (1976). They suggested that matrix secreted at a particular stage of crystal growth would be inhibitory and thus terminate the growth of the crystal layer. The free matrix surface that was formed would, however, then provide nucleation sites for the next crystal layer. This double action has been demonstrated by Addadi and Weiner (1985) with purified aspartic acid-rich protein from the soluble matrix of *Mytilus californianus*. Calcite crystals were inhibited from forming in solutions of this protein. If, however, the protein was coated on the walls of a glass vial containing a supersaturated solution, new crystals grew upon this surface. In this arrangement, the protein acted as a nucleator. The matrix undoubtedly has the same action in both situations, namely binding of calcium. If the matrix attached to the glass surface were released, we could expect that it would inhibit nucleation and crystal growth in solution. Serine-rich proteins did not show this inhibitory and nucleating capability, suggesting that the response required groups with negative charges. This conclusion was supported by the finding that the inhibitory properties of extrapallial fluid can be removed by passing it through an anion exchange column that binds negative groups (Wilbur and Bernhardt, 1984). Thus, it appears that the active groups of organic matrix are probably COO^- or OSO_3^- or both, and that these can attach to the calcium ions of the crystal lattice and cause inhibition of further crystal growth. If the secretion of this inhibitory organic matrix occurred at a particular stage of crystal growth, the thickness would be the same. Fracture surfaces of nacreous shell show this uniformity, as we mentioned. The nature of this precise control is yet to be explored.

These suggested explanations of nucleation and inhibition by the matrix leave unexplained the development of shell crystals during the period that they are surrounded by extrapallial fluid and soluble matrix. Two explanations of this lack of inhibition have been suggested. Crenshaw (personal communication) has found that crystal induction by soluble matrix covalently attached to a substrate required 2,000 times the concentration of the fixed protein to inhibit. That is, matrix in solution is not strongly inhibitory. Alternatively, the explanation may conceivably lie in the abolishing of inhibition by the binding by the soluble matrix of Ca^{2+} transported into the extrapallial fluid. Once all binding sites are occupied by calcium, the extrapallial fluid and matrix can no longer react with the nuclei and crystal lattices, thus permitting crystals to grow. Now the matrix may initiate the formation of nuclei and crystals; and crystal growth can continue to maximum thickness.

B. Prismatic Layer

The prismatic layer of molluscan shells is made up of long crystal prisms consisting of elongate spicular elements. The prisms extend inward from the periostracum and are surrounded by matrix that becomes sclerotized and separates individual prisms. The prisms of some species appear to have their origin in spherulites that develop on the inner surface of the newly formed sheet of periostracum (Taylor and Kennedy, 1969). It is from the crystallites of these spherulites that the long prisms are thought to form through growth in diameter and length. Longitudinal sections of prisms show the presence of irregularly spaced horizontal bands of organic material (Fig. 14.4a) Whether the growth of the prisms is inhibited at band regions is unknown. In any case, the crystal continues growth without change in diameter. This differs from the presumed inhibitory action of organic matrix in areas of lamellar shell.

The cause of the marked difference in crystal morphology in laminar and prismatic shell is a basic unanswered question of molluscan mineralization. At present, we can say only that the difference in crystal form is probably related to the organic matrix, known to be different in composition in the two types of shell.

VII. Endocrine Influences on Mineralization

The experiments of Lubet (1971) using classical extirpation and implantation methods on *Crepidula fornicata* suggested that the cerebral ganglia contained a factor that stimulated growth. This was confirmed by Geraerts (1976a) who demonstrated a growth factor in the light green cells of the cerebral ganglia of *Lymnaea stagnalis*. Removal of these cells caused a marked reduction in body and shell growth together with a decrease in oviposition and food intake. Levels of glucose and amino acids in the hemolymph, on the other hand, were increased. In view of these multiple effects, the mechanism of action of the factor in light green cells on shell formation was unclear. Moreover, the lateral lobes of the cerebral ganglia were found to have an inhibitory influence on the growth factor (Geraerts, 1976b, Roubos *et al.*, 1980).

To determine whether the light green cells had a more direct effect on mineral deposition, the cells were removed and changes in the shell edge and mantle edge examined. Three effects were observed: (1) a reduced rate of deposition of ^{47}Ca (Dogterom *et al.*, 1979) and $H^{14}CO_3$ (Dogterom and van der Schors, 1980) in the shell edge; (2) a reduction in the amount of a calcium binding complex in the mantle edge (Dogterom and Doderer, 1981); and (3) reduced incorporation of ^{3}H- tyrosine at the shell edge which was interpreted as a decreased formation of periostracum.[1] The effect on periostracum formation was considered crucial in inhibition of shell growth (Dogterom and Jentjens, 1980).

Kunigelis and Saleuddin (1978) also obtained evidence for a growth factor in the brain of *Helisoma duryi* by a different method. These investigators found that individuals of the same age in populations reared under constant conditions produced shell at different rates. If homogenates of the whole brain of snails with fast growing shells were injected into snails with slow growing shells, the rate of shell growth was substantially increased. The effectiveness of the homogenate was such that an amount equivalent to 1/1000 brain produced a detectable stimulation. Homogenates from supraesophageal regions were even more effective, suggesting the possibility of an inhibitory influence by other regions of the brain. Homogenates of brains of slow growers did not retard shell growth in fast growers.

Experiments were next carried out on effects of brain on periostracum formation by the mantle edge in physiological solution. This *in vitro* preparation has the advantage that the periostracum produced can be separated from the mantle and the amount measured. On addition of whole brain from snails with fast growing shell to isolated mantle from animals with slow-growing shell, the amount of periostracum produced was greater than on addition of whole brain from slow growing snails. The amount of periostracum was still further increased by removal of

[1] A correction for tyrosine incorporation in shell matrix was not carried out.

the lateral lobes before the brain was placed in culture (Kunigelis and Saleuddin, 1984). This retardation by the lateral lobes confirmed a comparable effect observed by Geraerts (1976b) in *Lymnaea* and by Geraerts and Mohamed (1981) in *Bulinus*.

Another molluscan mineralization system that may be found to be under hormonal influence is the reproductive system of terrestrial snails. As the eggs pass along the uterus to the exterior, crystals of $CaCO_3$ are deposited on their surface, and in some families a shell of considerable thickness is formed (Tompa, 1976). At the time of egglaying, increases occur in hemolymph calcium (Tompa and Wilbur, 1977) and protein (Kunigelis, personal communication). Also, the inward rate of calcium transport across the wall of the uterus as measured with ^{45}Ca, becomes greater. (Kunigelis, personal communication). These changes together with the known indirect influence of the light green cells on oviposition point to possible hormone action and suggest that experiments to examine hormone effects on ion transport may be worth consideration.

VIII. Additional Mineralizing Systems

A. Intracellular Calcium Deposits

We have previously noted that a large number of the biologically induced deposits of minerals are amorphous as shown by X-ray and electron diffraction (see Chapter 2). A number of such systems were discussed in the Protoctista (see Chapter 6) but the molluscs provide one of the best sources of such materials. There are two main types of such deposits (Simkiss and Mason, 1983). The first is found in the basophil cells of the digestive gland and consists of $CaMgP_2O_7$ in an amorphous form (Howard *et al.*, 1981). The pyrophosphate ion is able to bind a wide range of metal ions and the amorphous nature of these deposits means that the lattice structure is poorly developed and unlikely to exclude such foreign ions. These intracellular deposits accumulate a variety of metals from the diet of the mollusc and may be a way of detoxifying such materials (Simkiss, 1981). Calcium phosphate granules are also present in the gills of the freshwater bivalves *Anodonta grandis, Ligumia subrostrata* (Silverman *et al.*, 1988), and the inner shell surface of *Rangia cuneata* (Marsh, 1986).

The second type of molluscan intracellular deposit is, however, of more interest in this chapter. It occurs in the so-called connective tissue calcium cells widely distributed in molluscan tissues. Ultrastructural studies of these cells show that the granules are formed in membrane-bound vesicles of the Golgi—endoplasmic reticulum system, and although they are commonly in the amorphous state, crystals may be formed (Richardot and Wauthier, 1972). All granules thus far examined have an organic matrix containing protein. The calcium in these stores can be mobilized under various physiological conditions and pass out of the cells through pores which form between the plasma membrane and the membrane surrounding the granules (Sminia et al., 1977). Conditions of mobilization of granules are listed in Table 14.3. Watabe et al. (1976) demonstrated mobilization during shell repair and found that the calcium lost from the tissues was similar in amount to that deposited during repair (Table 14.4).

Another example of a somewhat different kind is the association of the extracellular concretions of freshwater unionids with reproduction. The concretions in the gills of these molluscs contain calcium and phosphate together with organic material. In Anodonta grandis the easily soluble and less soluble proteins extracted from the structures included 12–15 major proteins; glutamic, aspartic and glycine were the amino acids in highest concentration (Silverman et al., 1988). During the period that the animals are reproductively active, the concretions are mobilized (Silverman et al., 1985) and could provide calcium for shell formation of the glochidia (Silverman et al., 1987).

In instances in which mobilization and disappearance of concretions or granules are clearly correlated with a particular physiological function, one should be cautious in concluding that these structures have provided the only source of calcium. Other sources may also have contributed, particularly so since the body fluid compartments are undoubtedly interconnected.

The implications of the observations on these various calcareous storage bodies are sufficiently important that it is worth making three points

Table 14.3 Functions Involving Dissolution of Calcified Granules[a]

Formation of shell
Shell repair
Formation of epiphragm
Embryonic development
Buffer action against acidity of anaerobiosis

[a] Fournié and Chetail, 1982

Table 14.4 Evidence for the Depletion of Calcium from
Tissue Stores During Repair of the Shell of the Snail
Pomacea paludosa[a]

Tissue	Normal Snail (mg Ca/snail)	Snail with Shell Repair (mg Ca/snail)
Mantle	11.0	5.8
Foot	18.9	6.8
Rest of body	21.6	21.3
Total	51.5	33.9
Mass of repaired shell	—	20.4

[a] From Watabe *et al.* (1976).

about their interpretation. First, if there is, for example, a neurohormo-
nal control system in the coordination of eggshell formation, then it may
work by influencing resorption from the storage cells and calcium transport across the wall of the uterus. Second, the formation of amorphous
mineral deposits preceding the crystalline form (Richardot and
Wauthier, 1972) is a particularly interesting example of an organic matrix–mineral interaction. Clearly, in this case there is no close matching
of charges between the protein and inorganic spacing of the mineral.
Third, the energetics of the system is surprising. According to the
Ostwald–Lussac Law, minerals initially form in their most soluble and
least stable forms. This is because hydrated ions often have to be dehydrated in order to fit into a crystal lattice and this requires energy. Thus,
amorphous and somewhat hydrated minerals are often the first deposits
to form. It looks as if the connective tissue calcium cells accumulate
deposits in this form and then release them through pores in their membranes. Subsequently, this calcium is deposited in a crystalline form at
sites such as the mantle–shell or oviduct–eggshell locations.

The conditions that induce the formation of these intracellular granules and the mechanism of their dissolution are unsolved problems
relating to many phyla. Hopefully, they will be given experimental attention at the cellular and organism level.

B. The Radula

The story is told how Lowenstam was walking in the intertidal zone of
the beach on the tropical island of Palau when he noticed that some
limestone fossil reefs were eroded at their bases. The usual explanation,
that they were worn away by the sea, did not seem to fit with the fact

that they were most damaged as one moved toward the lagoon area. On closer examination, it became apparent that the limestone was being rasped away at the bottom of these rocks by chitons. This was surprising, however, since rocks are generally much harder than the protein and chitin tissues of molluscs and, on further investigation, Lowenstam

Figure 14.10 Median section through the buccal mass of *Lymnaea stagnalis*. Note the few odontoblasts (odg) at the lower posterior end of the radular sheath, the collostylar hood (ch) at the opening of the sheath, and the adhesive layer (al) which is limited to the functional part of the radula. In feeding, the cuticular radula is moved over the odontophor (o) by several muscles, which are concentrated in the buccal mass of the foregut. The radula is bent over the odontophor, which is located in the center of the buccal mass and bulges the functional part of the radula into the buccal cavity (bc). The odontophor is composed of two pieces which are connected by muscles and they together form a trough into the cavity. During feeding, the odontophor is turned vertically so that the front edge (fe) touches the nutriment. The radula is not intimately connected with the odontophor and can be moved to and fro by the supramedian tensor muscle (tsm) or the laterally located protractors. The posterior part of the radula is located in a blind pouch, the radular sac or radular sheath. It is a U-shaped groove and the lateral teeth are directed inward. The collostyle (col) protrudes into the groove from above. bce, buccal cavity epithelium; che, collostylar hood-epithelium; dre, distal radular epithelium; dz, degeneration zone; ie, inferior epithelium; j, jaw; mit, mitosis; rm, radular membrane; rt, radula teeth; se, superior epithelium. (Mackenstedt and Märkel, 1987).

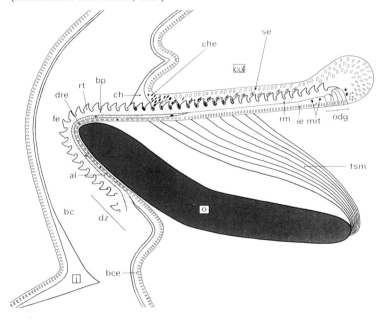

found that the rasping tongue or radula was impregnated with magnetite (Kirschvink *et al.*, 1985).

In the 25 years since this discovery, biogenic magnetite has attracted increasing attention. What was once considered to be a biochemical oddity is now recognized as a form of biomineralization that occurs in other phyla as well. The radula remains, however, as one of the best studied systems.

The radula of chitons consists of a long ribbon, secreted in a sheath of the ectodermal foregut and bearing on its surface a number of rows of

Figure 14.11 Scanning electron micrographs of the unworn portion of radula of *Urosalpinx cinerea follyensis*. (a) Rachidian teeth, intermediate grooves, and marginal teeth at bending plane, ×200; (b) unworn rachidian teeth illustrating sharpness of the five major cusps, two small cusps, denticles, and the interlocking of adjacent teeth (arrow), ×575; (c) side view of rachidian cusps and denticles illustrating recurved nature, especially of major central cusp, ×980; (d) side view of marginal teeth illustrating their sharp, hooked (scythe-shaped) ends. ×980 (Carriker *et al.*, 1974, *Mar. Biol.* **25**, 63–76). (Reprinted by permission of Springer-Verlag.)

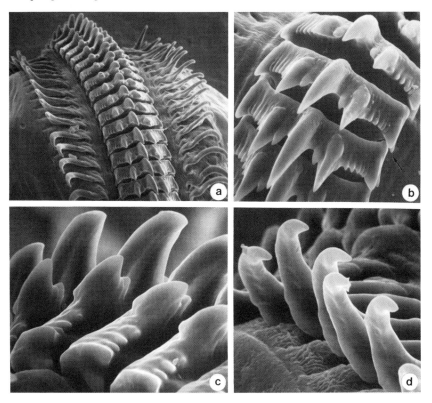

teeth (Figs. 14.10 and 14.11). The caps of the major lateral teeth are reinforced with crystalline magnetite in an ordered matrix of organic fibrils. The biomineralization process proceeds in two steps (Towe and Lowenstam, 1967). There is an initial deposit of ferrihydrite ($5Fe_2O_3 \cdot 9H_2O$) which is then converted to magnetite ($FeO \cdot Fe_2O_3$). The epithelium of the radula sac over this site of mineralization contains considerable quantities of ferritin which is thought to provide a source of iron. The cells overlying the tooth caps have extensive microvilli and the crystals of magnetite that are deposited appear to lie at specific angles to the long axis. The spacing and orientation of the mineral on the protein–chitin fibrils of the radula suggest a role for the organic matrix in the nucleation and growth of the crystals (Nesson and Lowenstam, 1985).

The radulas of other molluscs, for example the limpet *Patella vulgata*, are also mineralized but contain about 10% silica as hydrated opal and 12% ferric oxide as goethite ($FeOOH$). Mineralization of a tooth cusp in these animals begins with goethite impregnation at the posterior region and consists of thin strands of mineral 15–20 nm wide) arranged with their 001 crystallographic axis parallel to the tooth wall. The process continues with an increase in the number and thickness of the crystals. The mature crystals are well ordered but with extensive growth distortions attributed to the filamentous organic matrix. According to Mann *et al.* (1986), silica impregnates the matrix at a later stage in mineralization. This confirms the earlier work of Runham *et al.* (1969) that suggested that the iron oxide and silica phases were spatially separated. It seems to be generally agreed, therefore, that silica infiltrates the radula and fills spaces left between the goethite crystals. Such composites are rare in biomineralization and deserve further study.

The formation of the radula in pulmonate snails has been described by Mackenstedt and Märkel (1987). The most posterior cells or odontoblasts first secrete the matrix of the cusp and distal tooth surface (Fig. 14.10). The entire ribbon of teeth then progresses anteriorly so that the next odontoblasts on coming in contact with the forming tooth secrete more matrix and enlarge the tooth. As the developing radula moves anteriorly, the base is added by other cells until a final shaping of the matrix takes place. Thus, odontoblasts in linear arrangement demonstrate a marvellously integrated sequence. Not only do different cells secrete specific amounts of matrix for particular portions of the tooth but they shape it as well. To finish the process, another group of cells, the superior epithelium, adds ions and organic compounds, which results in hardening (Fig. 14.11).

The radula is destroyed at its anterior end and is constantly replaced by the newly formed regions as they advance anteriorly. Moreover, the rates of disintegration and formation are balanced. This could come about as the newly synthesized portions are disintegrated on reaching particular cells of the distal radular epithelium (Fig. 14.10). By labeling the teeth with ^{35}S, Mackenstedt and Märkel (1987) determined the time for complete replacement of radulae. For *Cepaea nemoralis* the period was 30–35 days and for *Lymnaea stagnalis* 24 days.

IX. Summary

1. Molluscs carry out a remarkable diversity of mineralizing activities. They form more than 24 mineralized structures involving some 20 minerals.
2. The greater number of these structures is formed by a mineralizing epithelium, the cells of which may deposit more than one mineral and more than one crystal arrangement.
3. Molluscan shell formation involves transport of Ca^{2+} and HCO_3^- across the outer mantle epithelium and the secretion of an organic matrix. As a result, the inner surface of the shell increases in thickness incrementally. The mechanism of this incremental deposition is unknown.
4. The microstructure of shells of various species represents an impressive array of crystal forms in well-defined crystal arrangements. Presumably, their form and arrangement are controlled by the three-dimensional organization of the organic material secreted by the epithelium and within which the crystals develop.
5. Neurohumoral substances of the ganglia of snails affect the growth rate of the shell.
6. Molluscs form intracellular deposits that are usually amorphous and raise interesting questions about their chemical induction, mobilization, and biological functions.
7. The radula of chitons and patellid gastropods is reinforced with metals in the formation of hard rasping teeth. The formation of the teeth, the role of iron deposits in this process, and the control of crystal deposition by the matrix and radula sheath cells present novel aspects of biomineralization.

References

Addadi, L., and Weiner, S. (1985). Interactions between acidic proteins and crystals: Stereochemical requirements in biomineralization. *Proc. Natl. Acad. Sci. U.S.A.* **82,** 4110–4114.

Bandel, K. (1979). Übergänge von einfacheren Strukturtypen zur Kreuzlamel-lenstruktur bei Gastropodenschalen. *Biomineralisation* **10,** 9–38.

Borle, A. B., and Snowdowne, K. W. (1982). Measurement of intracellular free calcium in monkey kidney cells with aequorin. *Science* **217,** 252–254.

Campbell, J. W., and Speeg, K. V. (1969). Ammonia and the biological deposition of calcium carbonate. *Nature (London)* **224,** 725–726.

Carriker, M. R., Schaadt, J. G., and Peters, V. (1974). Analysis of slow-motion picture photography and scanning electron microscopy of radular function in *Urosalpinx cincreafollyensis* (Muricidae, Gastropoda) during shell penetration. *Mar. Biol.* **25,** 63–76.

Carter, J. G., and Clark, G. R. (1985). Classification and phylogenetic significance of molluscan shell microstructure. *In* "Mollusks. Notes for a Short Course" (T. W. Broadleaf, ed.), Dept. of Geol. Sci. Stud. Geol. 13, pp. 50–71. Univ. of Tennessee, Knoxville, Tenn.

Clarkson, D. T., Brownlee, C., and Ayling, S. M. (1988). Cytoplasmic calcium measurements in intact higher plant cells: results from fluorescence ratio imagings of fura-2. *J. Cell Sci.* **91,** 71–80.

Coimbra, J., Machado, J., and Fernandes, P. L. (1988). Electrophysiology of the mantle of *Anodonta cygnea. J. Exp. Biol.* **140,** 65–88.

Crenshaw, M. A. (1972). The inorganic composition of molluscan extrapallial fluid. *Biol. Bull.* **143,** 506–512.

Crenshaw, M. A. (1980). Mechanisms of shell formation and dissolution. *In* "Skeletal Growth of Aquatic Organisms" (D. C. Rhoads and R. A. Lutz, eds.), pp. 115–132. Plenum, New York.

Crenshaw, M. A., and Neff, J. M. (1969). Decalcification at the mantle-shell interface in molluscs. *Am. Zool.* **9,** 881–885.

Crenshaw, M. A., and Ristedt, H. (1976). The histochemical localization of reactive groups in septal nacre from *Nautilus pompilus. In* "The Mechanisms of Mineralization in the Invertebrates and Plants" (N. Watabe and K. M. Wilbur, eds.), pp. 355–367. Univ. of South Carolina Press, Columbia, South Carolina.

Dillaman, R. M., and Ford, S. E. (1982). Measurement of calcium carbonate deposition in molluscs by controlled etching of radioactively labelled shells. *Mar. Biol.* **66,** 133–143.

Dipolo, R., Requena, J., Brinley, F. J., Jr., Mullins, L. J., Scarpa, A., and Tiffert, T. (1976). Ionized calcium concentrations in squid axons. *J. Can. Physiol.* **67,** 433–467.

Dogterom, A. A. and vander Schors, R. C. (1980). The effect of the growth hormone of *Lymnaea stagnalis* on (bi) carbonate movements, especially with regard to shell formation. *Gen. Comp. Endocrinol.* **41,** 334–339.

Dogterom, A. A., and Doderer, A. (1981). A hormone dependent calcium-binding protein in the mantle edge of the freshwater snail *Lynmaea stagnalis. Calcif. Tissue Res.* **33,** 505–508.

Dogterom, A. A., and Jentjens, T. (1980). The effect of the growth hormone of the snail *Lymnaea stagnalis* on periostracum formation. *Comp. Biochem. Physiol. A* **66,** 689–690.

Dogterom, A. A., van Loenhout, H., and van der Schors, R. C. (1979). The effect of the growth hormone of *Lymnaea stagnalis* on shell calcification. *Gen. Comp. Endocrinol.* **39,** 63–68.

Enyikwola, O., and Burton, R. F. (1983). Chloride-dependent electrical potentials across the mantle epithelium of *Helix. Comp. Biochem. Physiol. A* **74,** 161–164.

Fournié, J. and Chétail, M. (1982). Accumulation calcique au niveau cellulaires chez les mollusques. *Malacol.* **22,** 265–284.

Geraerts, W. P. M. (1976a). Control of growth by the neurosecretory hormone of the light green cells of the freshwater snail *Lymnaea stagnalis. Gen. Comp. Endocrinol.* **29,** 61–71.

Geraerts, W. P. M. (1976b). The role of the lateral lobes in the control of growth and reproduction in the hermaphrodite freshwater snail *Lymnaea stagnalis. Gen. Comp. Endocrinol.* **29,** 97–102.

Geraerts, W. P. M. and Mohamed, A. M. (1981). Studies on the role of the lateral lobes and the ovotestis of the pulmonate snail *Bulinus truncatus* in the control of body growth and reproduction. *Int. J. Invertebr. Reprod.* **3,** 297–308.

Gordon, J. and Carriker, M. R. (1980). Sclerotized protein in the shell matrix of a bivalve mollusc. *Mar. Biol.* **57,** 251–260.

Greenaway, P. (1971). Calcium regulation in the freshwater mollusc *Lymnaea stagnalis* (L.) (Gastropoda: Pulmonata). II. Calcium movements between internal calcium compartments. *J. Exp. Biol.* **54,** 609–620.

Howard, B., Mitchell, P. C. H., Ritchie, A., Simkiss, K., and Taylor, M. (1981). The composition of intracellular granules from the metal-accumulating cells of the common garden snail (*Helix aspersa*). *Biochem. J.* **194,** 507–511.

Istin, M., and Kirschner, L. B. (1968). On the origin of the bioelectrical potential generated by the freshwater clam. *J. Gen. Physiol.* **51,** 478–496.

Jodrey, L. H. (1953). Studies on shell formation. III. Measurements of calcium deposition in shell and calcium turnover in mantle tissue using the mantle-shell preparation and Ca^{45}. *Biol. Bull. (Woods Hole, Mass.)* **104,** 398–407.

Kirschvink, J. L., Jones, D. S., and MacFadden, B. J. (eds.) (1985). "Magnetite Biomineralization and Magnetoreception in Organisms," pp. 1–682. Plenum, New York.

Krampitz, G. and Witt, W. (1979). Biomineralization. In "Topics in Current Chemistry, Biochemistry," Vol. 78 (F. L. Boschke, ed.), pp. 57–144. Springer Verlag, Berlin.

Kunigelis, S. C., and Saleuddin, A. S. M. (1978). Regulation of shell growth in the pulmonate gastropod *Helisoma duryi. Can. J. Zool.* **56,** 1975–1980.

Lowenstam, H. A., and Weiner, S. (1983). Mineralization by organisms and the evolution of biomineralization. *In* "Biomineralization and Biological Metal Accumulation" (P. Westbroek and E. W. de Jong, eds.), pp. 191–203. Riedel, Dordrecht, The Netherlands.

Lubet, P. (1971). Influence des ganglions cerebroides sur la croissance de *Crepidula fornicata* Phil. (Mollusque Mesogastecopode). *C.R. Acad. Sci.* Paris **273,** 2309–2311.

Mackenstedt, U., and Märkel, K. (1987). Experimental and comparative morphology of radula renewal in pulmonates (Mollusca, Gastropoda). *Zoomorphology* **107,** 209–239.

Mann, S. (1983). Mineralization in biological systems. *Struct. Bonding (Berlin)* **54,** 125–174.

Mann, S., Perry, C. C., Webb, J., Luke, B., and Williams, R. J. P. (1986). Structure, morphology, composition, and organization of biogenic minerals in limpet teeth. *Proc. R. Soc. London, Ser. B.* **227,** 179–190.

Manyak, D. M. (1982). "Calcified Tube Formation by the Shipworm *Bankia gouldi*," Ph.D. thesis. Duke University, Durham, North Carolina.

Marsh, M. E. (1986). Biomineralization in the presence of calcium-binding phosphorpro-
tein particles. *J. Exp. Zool.* **239**, 207–220.

Meenakshi, V. R., Martin, A. W., and Wilbur, K. M. (1974). Shell repair in *Nautilus
macrophalus*. *Mar. Biol.* **27**, 27–35.

Meenakshi, V. R., Hare, P. E., and Wilbur, K. M. (1971). Amino acids of the organic matrix
of neogastropod shells. *Comp. Biochem. Physiol.* **40B**, 1037–1043.

Neff, J. M. (1972). Ultrastructure of the outer epithelium of the mantle in the clam *Mercenaria mercenaria* in relation to calcification of the shell. *Tissue Cell* **4**, 591–600.

Nesson, M. H., and Lowenstam, H. A. (1985). Biomineralization processes of the radular
teeth of chitons. *Top. Geobiol.* **5**, 333–363.

Ravindranath, M. H., and Rajeswari-Ravindranath, M. H. (1974). The chemical nature of
the shell of molluscs: I. Prismatic and nacreous layers of a bivalve *Lamellidans margina-
lis* (Unionidae). *Acta Histochem.* **48**, 26–41.

Richardot, M., and Wauthier, J. (1972). Les cellules à calcium du tissu conjonctif de
Ferrissia wauthieri (Mirolli). Données écologiques, biologiques et physiologiques.
Thèse, Lyon, n° 76–6.

Roubos, E. W., Geraerts, W. P. M., Boerrigter, G. H., and van Kampen, G. P. J. (1980).
Control of the activities of the neurosecretory light green and caudo-dorsal cells and
the endocrine dorsal bodies by the lateral lobes in the freshwater snail *Lymnaea stagnalis* (L.). *Gen. Comp. Endocrinol.* **40**, 446–454.

Runham, N. W., Thornton, P. R., Shaw, D. A., and Wayte, R. C. (1969). The mineralization and hardness of the radular teeth of the limpet *Patella vulgata* L. *Z. Zellforsch.* **99**,
608–626.

Samata, T., Sanguansri, P., Cazaux, C., Hamm, M., Engels, J., and Krampitz, G. (1980).
Biochemical studies on components of mollusc shells. *In* "The Mechanisms of
Biomineralization in Animals and Plants" (M. Omori and N. Watabe, eds.), pp. 37–
47. Tokai Univ. Press, Tokyo.

Sikes, C. S. and Wheeler, A. P. (1983). A systematic approach to some fundamental
questions of carbonate calcification. *In* "Biomineralization and Biological Metal Accumulation" (P. Westbroek and E. W. de Jong, eds.), pp. 285–290. Reidel, Dordrecht,
Holland.

Silverman, H., Steffens, W. L., and Dietz, T. H. (1985). Calcium from extracellular concretions in the gills of freshwater unionid mussels is mobilized during reproduction. *J.
Exp. Zool.* **236**, 137–147.

Silverman, H., Kays, W. T., and Dietz, T. H. (1987). Maternal calcium contribution to
glochidial shells in freshwater mussels (Eulamellibranchia: Unionidae). *J. Exp. Zool.*
242, 137–146.

Silverman, H., Sibley, L. D., and Steffens, W. L. (1988). Calmodulin-like calcium-binding
protein identified in calcium-rich mineral deposits from freshwater mussel gills.
J. Exp. Zool. **247**, 224–231.

Simkiss, K. (1976). Cellular aspects of calcification. *In* "The Mechanisms of Mineralization
in the Invertebrates and Plants" (N. Watabe and K. M. Wilbur, eds.), pp. 1–31. Univ.
of South Carolina Press, Columbia, South Carolina.

Simkiss, K. (1981). Cellular discrimination processes in metal accumulating cells. *J. Exp.
Biol.* **94**, 317–327.

Simkiss, K. (1984). The karyotic mineralization window (KMW). *Am. Zool.* **24**, 847–856.

Simkiss, K., and Mason, A. Z. (1983). Metal ions: Metabolic and toxic effects. *In* "The
Mollusca" (P. Hochachka, ed.), Vol. 2, pp. 101–164. Academic Press, New York.

Sminia, T., de With, N. D., Bos, J. L., Nieuwmegen, M. E., Witter, M. P., and Wondergem, J. (1977). Structure and function of the calcium cells of the freshwater pulmonate snail, *Lymnaea stagnalis*. *Neth. J. Zool.* **27**, 195–208.

Suzuki, S. and Uozumi, S. (1981). Components of prismatic layers of molluscan shells. *J. Fac. Sci. Hokkaido Univ.* Ser IV **20**, 7–20.

Taylor, J. D. (1973). The structural evolution of the bivalve shell. *Palaeontology* **16**, 519–534.

Taylor, J. D., and Kennedy, W. J. (1969). The influence of the periostracum on shell structure of bivalve molluscs. *Calcif. Tissue Res.* **3**, 274–283.

Tompa, A. (1976). Embryonic use of eggshell calcium in a gastropod. *Nature, Lond.* **255**, 232–233.

Tompa, A. S., and Wilbur, K. M. (1977). Calcium mobilization during reproduction in the snail, *Helix aspersa*. *Nature (London)* **270**, 53–54.

Towe, K. M., and Lowenstam, H. A. (1967). Ultrastructure and development of iron mineralization in the radular teeth of *Cryptochiton stelleri* (Mollusca). *J. Ultrastructure Res.* **17**, 1–13.

Watabe, N. (1965). Studies on shell formation. XI. Crystal-matrix relationships in the inner layers of mollusk shells. *J. Ultrastruc. Res.* **12**, 351–370.

Watabe, N. (1981). Crystal growth of calcium carbonate in the invertebrates. *Prog. Cryst. Growth Charact.* **4**, 99–147.

Watabe, N. (1984). Shell. *In* "Biology of the Integument. I. Invertebrates" (J. Bereiter-Hahn, A. G. Matoltsy, and K. S. Richards, eds.), pp. 448–485. Springer-Verlag, Berlin.

Watabe, N. (1988). Shell structure. *In* "The Mollusca" (E. R. Trueman and M. R. Clarke, eds.), Vol. 11, pp. 69–104. Academic Press, San Diego, California.

Watabe, N., Meenakshi, V. R., Blackwelder, P. L., Kurtz, E. R., and Dunkleberger, D. G. (1976). Calcareous spheres in the gastropod *Pomacea paludosa*. *In* "The Mechanisms of Mineralization in the Invertebrates and Plants" (N. Watabe, and K. M. Wilbur, eds.), pp. 283–308. Univ. of South Carolina Press, Columbia, South Carolina.

Weiner, S. (1983). Mollusk shell formation: Isolation of two organic matrix proteins associated with calcite deposition in the bivalve, *Mytilus californianus*. *Biochemistry* **22**, 4139–4144.

Weiner, S., and Traub, W. (1980). X-ray diffraction study of the insoluble matrix of mollusc shells. *FEBS Lett.* **111**, 311–316.

Weiner, S., and Traub, W. (1981). Organic–matrix–mineral relationshipos in mollusk shell nacreous layers. *In* "Structural Aspects of Recognition and Assembly in Biological Macromolecules" (M. Balaban, J. L. Sussman, W. Traub, and A. Yonath, eds.), pp. 467–482. Balaban ISS, Rehovet, Israel.

Wheeler, A. P. (1975). "Oyster Carbonic Anhydrase: Evidence for Plasma Membrane-Bound Activity and for a Role in Bicarbonate Transport," Ph.D. thesis. Duke University, Durham, North Carolina.

Wheeler, A. P., Blackwelder, P. L., and Wilbur, K. M. (1975). Shell growth in the scallop *Argopecten irradians* Isotope incorporation with reference to diurnal growth. *Biol. Bull. (Woods Hole, Mass.)* **148**, 472–482.

Wheeler, A. P., George, J. W., and Evans, C. A. (1981). Control of calcium carbonate nucleation and crystal growth by the soluble matrix of oyster shell. *Science* **212**, 1397–1398.

Wheeler, A. P., Rusenko, K. W. and Sikes, C. S. (1988). Organic matrix from carbonate biomineral as a regulator of mineralization. In "Chemical Aspects of Regulation of

Mineralization" (C. S. Sikes and A. P. Wheeler, eds.), pp. 9–13. Univ. South. Alabama Publ. Serv., Mobile, Alabama.

Wilbur, K. M. (1972). Shell formation in molluscs. *Chem. Zool.* **7,** 103–145.

Wilbur, K. M. (1984). Many minerals, several phyla, and a few considerations. *Am. Zool.* **24,** 839–845.

Wilbur, K. M. (1985). Topics in molluscan mineralization: Present status, future directions. *Am. Malacol. Bull., Spec. Ed.* **1,** 51–58.

Wilbur, K. M., and Bernhardt, A. M. (1984). Effects of amino acids, magnesium, and molluscan extrapallial fluid on crystallization of calcium carbonate: *In vitro* experiments. *Biol. Bull. (Woods Hole, Mass.)* **166,** 251–259.

Wilbur, K. M., and Saleuddin, A. S. M. (1983). Shell formation. *In* "The Mollusca," Vol. 4 (A. S. M. Saleuddin and K. M. Wilbur, eds.), pp. 235–287. Academic Press, New York.

Wilbur, K. M., and Simkiss, K. (1968). Calcified shells. *Compr. Biochem.* **26A,** 229–295.

Wilkes, D. A. and Crenshaw, M. A. (1979). Formation of a dissolution layer in molluscan shells. In "Scanning Electron Microscopy/1979/II" (O. Johari and R. P. Becker, eds.), pp. 469–474. SEM, O'Hare, Illinois.

Wise, S. W., Jr. (1970). Microarchitecture and mode of formation of nacre (mother-of-pearl) in pelecypods, gastropods, and cephalopods. *Eclogae Geol. Helv.* **63,** 775–797.

Zischke, J. A., Watabe, N., and Wilbur, K. M. (1970). Studies on shell formation: Measurements of growth in the gastropod *Ampullarius glaucus. Malacologia* **10,** 423–439.

15

Brachiopods—
Fluorapatites and
Calcareous Shells

The brachiopods are of special interest for biomineralization studies in that some genera form shells of calcium carbonate whereas others deposit shells of calcium phosphate. In fact, the division of the phylum into two classes follows these two mineral types. The class Articulata includes genera with calcareous shells and the genera of the class Inarticulata have phosphatic shells. The exception is the superfamily Craniacea, placed taxonomically with the Inarticulata but with shells of calcium carbonate. All brachiopods are marine. Although more than 1600 brachiopod genera have been described, there are now only some 70 living genera (Williams, 1965). For a detailed account of the brachiopods, the two volumes edited by Moore (1965) provide an excellent source of information.

Members of the phylum Brachiopoda or lampshells (because of their resemblance to ancient oil lamps) have two valves, and in this respect they bear a superficial resemblance to bivalve molluscs, although there is no hinge joining the valves in the Inarticulata. However, two brachiopod organs are distinct. One is the lophophore, a large ciliated feeding organ located anteriorly within the cavity formed by the two valves (Fig. 15.1). The other is the fleshy pedicle of the posterior region which extends out of the valves and anchors the animal to the substratum. The relationships of this ancient group of animals are obscure. They were abundant 600 million years ago but are now only sporadically distributed around the world. It is thought that they may be survivors of the

Figure 15.1 Section through an articulate brachiopod. (From Williams, A., and Rowell, A. J., 1965: *In* Moore, R. C. (ed.): *Treatise on Invertebrate Paleontology.* Geological Society of America and the University of Kansas.)

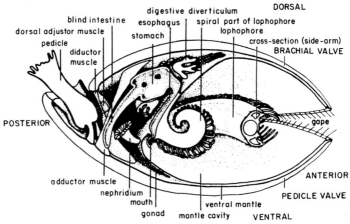

early stock of animals (deuterostomes) that was eventually to evolve into the vertebrates. However, the evidence for this lineage is problematical.

I. The Mineral System

The two valves of brachiopods, like those of molluscs, are formed by the outer epithelium of the mantle. Figure 15.2 shows the relationship of the mantle and shell in the articulate *Notosaria*. The mantle morphology of other genera may be somewhat different, although the general pattern is similar. The mineral layers, commonly of two structural types, are deposited by different mantle areas; and the periostracum, which covers the outer shell surface, is elaborated by a groove on the mantle surface near the growing edge.

Shells of Articulata and Inarticulata differ not only in the type of mineral deposited but also in the organic material secreted by the mantle epithelium during mineral deposition (Jope, 1977, 1979; Fig. 15.3). The shell protein content is strikingly different in the two classes. Whereas articulate calcareous shells have a protein content of about 0.5%, inarticulate phosphatic shells may have up to 25% protein, and the protein is associated with chitin. The high organic content is clearly evident from the relatively thick layers of organic material separating the mineral

Figure 15.2 Diagrammatic sagital section of the edge of a valve of *Notosaria*. (After Williams, 1977.)

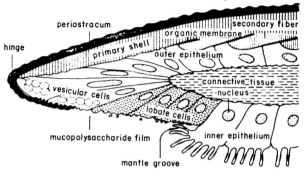

layers seen in sections of phosphatic valves. This alternating layered arrangement of mineral and organic material is not found in articulate shells. The amino acid composition of the protein of the two classes is also different. Articulate shell protein has a high glycine content and lacks hydroxyproline, except in *Crania*, an inarticulate with a calcareous shell. In the inarticulate *Lingula*, alanine and glycine account for more

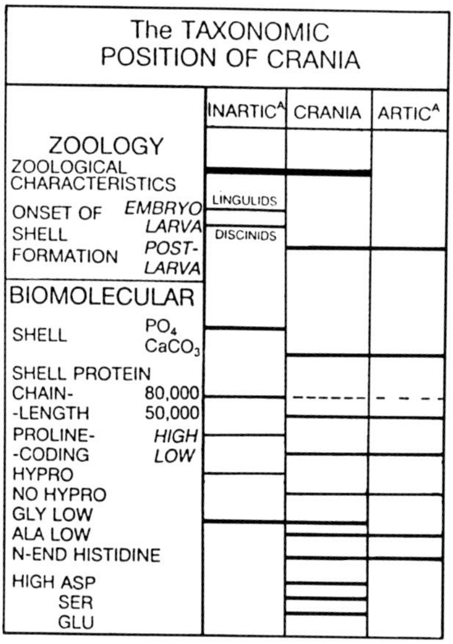

The TAXONOMIC POSITION OF CRANIA			
	INARTIC^A	CRANIA	ARTIC^A
ZOOLOGY			
ZOOLOGICAL CHARACTERISTICS			
ONSET OF SHELL FORMATION EMBRYO LARVA POST-LARVA	LINGULIDS DISCINIDS		
BIOMOLECULAR			
SHELL PO₄ CaCO₃			
SHELL PROTEIN CHAIN--LENGTH 80,000 50,000			
PROLINE--CODING HIGH LOW			
HYPRO NO HYPRO			
GLY LOW			
ALA LOW			
N-END HISTIDINE			
HIGH ASP SER GLU			

Figure 15.3 Diagram summarizing the anomalous taxonomic position of *Crania*, zoologically placed with the Inarticulata, but on most biomolecular criteria running with the Articulata. Most striking is the high *total* coded proline (i.e., proline + hydroxyproline) of inarticulates (except *Crania*) of 11.0–19.2, as compared with the low coded proline of articulates of 1.9–8.8 and the low coded proline of *Crania* 4.9–6.6 (Jope, 1977, with additions).

than one-third of the total (Iwata, 1981). Hydroxyproline was present but hydroxylysine was not detected. The presence of hydroxyproline cannot be taken as evidence of collagen, however, since the low glycine content in some genera is inadequate to form collagen molecules (Jope, 1977). Other differences in amino acids of shell proteins of the two classes are indicated in Fig. 15.3.

The proteins of the apatite shell of *Lingula* can be separated into three fractions: (1) an acid-soluble fraction, (2) a second acidic fraction associated with mineral and released on demineralization with EDTA; and (3) a hydrophobic residue that is the major protein fraction. Fractions (1) and (2) are both complex mixtures of proteins differing in their amino acid contents and electrophoretic patterns (Tuross, 1986). Tanaka *et al.* (1988) have purified a Ca-binding protein fraction called apslin from *Lingula* shell. It has a mass of 140 kDa, high concentrations of glycine, aspartic acid–asparagine and glutamic acid–glutamine, and is strongly adsorbed to hydroxyapatite crystals. The location of this protein within the shell has yet to be determined.

II. Mechanisms of Mineralization

A. Articulata

Notosaria, which has a two-layered shell of calcite, provides a useful example of shell formation in the Articulata (Williams, 1971; 1977; Gaspard, 1978). As we mentioned, the outer epithelium is directly responsible for mineral deposition and periostracum formation as in molluscs. The outer shell layer, called the primary layer (Fig. 15.2), is in contact with the inner periostracal surface and is made up of crystallites inclined at an acute angle to the outer face of the secondary layer surface below (not shown). The outer ends of the crystallite units can be seen on stripping off the periostracum and are probably nucleated on the inner periostracal surface. The units are also clearly seen in the primary layer of the calcitic inarticulate *Crania*.

The secondary layer of the valve is distinct in its structure from that of the primary layer and is formed by a part of the mantle epithelium further from the mantle tip than the portion forming the primary layer (Fig. 15.2). In this respect, shell deposition is similar to that in a bivalve that deposits two types of crystal layers. However, differences in shell ultrastructure in most brachiopods and molluscs are marked. Williams

(1971) has described a most remarkable phenomenon in the secondary layer which consists of elongate calcite fibers, each formed and shaped by a single cell of the mantle epithelium (Fig. 15.2). These mineral fibers and, in fact, most other features of the adult shell are also evident in the juvenile shell within a few days after metamorphosis (Stricker and Reed, 1985) (Fig. 15.4a,b). The growing end of the mineral fiber is in contact with the cell by which it is formed, and continuing deposition at the outer cell surface elongates the fiber throughout the period that the cell remains active. The fiber is sheathed in material secreted both by the cell from which it originates and by neighboring cells. Fiber elongation must require the intake of calcium by the free portions of the cell membrane and an active pumping of calcium and bicarbonate ions through the cell membrane to the site of deposition. This unusual method of mineralization is roughly comparable to that of extracellular mineralization by single cells of Porifera and Echinodermata. It differs from the indirect secretion of shell crystals which form from the extrapallial fluid, although this is the method that forms the primary layer of articulates and both layers of inarticulate shells.

B. Inarticulata

In *Crania,* an inarticulate brachiopod with a calcitic shell, the secondary layer is formed differently. Crystal tablets develop through screw dislocations into laminae. The mineral laminae are separated by layers of protein, and both mineral and protein are derived from the extrapallial fluid that is secreted by the mantle cells.

An examination of tangential sections of more typical Inarticulata valves clearly shows structural differences from Articulata and among genera of Inarticulata. We shall give very brief accounts of shell microstructure of two species that show marked differences: *Lingula unguis* (Iwata, 1981) and *Glottidia pyramidata* (Watabe and Pan, 1984; Pan, 1985).

The mineral of inarticulates is a calcium phosphate which is a CO_3 + F-containing calcium-OH-apatite (LeGeros *et al.*, 1985), similar (McConnell, 1963; Iwata, 1981) but not identical to francolite (LeGeros *et al.*, 1985). In the mineral layers of the valves of *Lingula* and *Glottidia*, the fluoride content averaged $2.5 \pm 0.37\%$ (Pan, 1985).

The clear-cut division of a shell of an articulate brachiopods into primary and secondary layers is not evident in *Lingula*. Rather, the shell consists of mineral layers (Fig. 15.5a), each pair separated by an organic layer called a chitin layer (Fig. 15.5b). As the shell increases in thickness, the alternation of mineral and organic layers is repeated, and, in a

Figure 15.4 (a) The inner side of a pedicle valve of *Terebratalia transversa* at 23 days postmetamorphosis. The asterisk marks the site of a poorly developed muscle scar. p, Puncta; t, tooth. Scale bar = 50 μm. (b) Edge of a pedicle valve 23 days postmetamorphosis. Asterisk marks border between primary shell on the right and the fibrous secondary shell on the left. mk, Median keel of shell fiber; sf, shell fiber of secondary layer. Scale bar = 10 μm (Stricker and Reed, 1985).

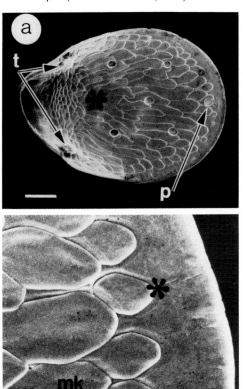

Lingula shell of 3.5 cm length, 16–17 of these double units will have been formed (Iwata, 1981). Each mineral layer is usually divided into three zones (Fig. 15.5b), although the C zone is not always present. The C zone, which is the first formed region, consists of grains and crystallites ranging from less than 10 nm to ~50 nm. Needle crystallites in the A zone are larger, with those in the B zone larger still. Deposition of the B zone is followed by the formation of an organic layer. The sequence then repeats.

Glottidia valves have two major layers, primary and secondary, unlike those of the articulate *Notosaria*, previously described. The primary layer is formed by columnar epithelial cells in the region of the outer mantle edge, and the secondary layer is deposited by two types of squamous cells of the general outer epithelium. Spherulites embedded in organic matrix make up the primary layer (Fig. 15.6), their density of packing decreasing with increasing distance from the periostracum. This decrease in density is also evident in X-ray microanalyses which indicate that calcium and phosphate are almost absent at the inner border of the layer (Pan, 1985). The secondary layer is made up of layers of spherulites fused into solid sheets. On the upper and lower surfaces of these layers large spherulites are present. Also, aggregated spherulites in the form of rods occur within the chitin layers separating the mineral layers (Pan and Watabe, 1988a).

Within the primary and secondary mineral layers at least two types of organic matrices have been identified. One consists of protein fibers 5 nm in diameter. The other is made of glycosaminoglycan in the form of a network of 5-nm fibers that surround and fill the interstices of the spherulites. The relationship of these two components has not been resolved but the protein appears to be at the periphery of the glycosaminoglycan networks.

Williams (1970) has described a different type of layer formation for the brachial valve of *Crania*. In this genus, the secondary shell is made up of layers of calcite tablets separated by protein layers. The tablets grow laterally by screw dislocation on a protein sheet forming part of a helicoid spiral. Another protein sheet then forms on the surface of the tablets. The sheets become continuous with one another and usually pass from one level to the next. The result is a spiral arrangement of alternating organic and calcite layers. It is stated that "Lateral accretion along a spirally-developing edge, which is the mode of growth in material affected by screw dislocation, is a continuous and *not* an episodic process." If the statement is taken to mean that there is no interruption in the growth process, *Crania* would differ in its skeletal development

Figure 15.5   (a) Low-magnification image of alternation of mineral (broad strips) and organic material of *Lingula unguis*. The two vertical lines are line profile by electron microprobe analysis (Iwata, 1981). (b) Diagram showing zones within a single minera-

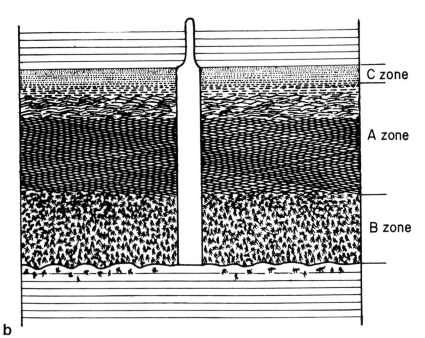

lized layer of *Lingula*. The horizontal lines at the top and bottom indicate organic layers that separate mineral layers. The mineralized layer may consist of A and B zones only. C zone, when present, is always outermost. A zone is the major layer and consists of needlelike crystallites, 1000–1500 Å in length, arranged subparallel to the shell surface or inclined at a very low angle. B zone consists of crystallites somewhat longer and larger in cross section and made up of thinner crystallites. They are deposited irregularly. C zone consists of amorphous or crystalline particles from less than 100 to about 500 Å in diameter. A single tubule or puncta is shown passing through the layer (Iwata, 1981).

Figure 15.6 Spherulites of the primary layer of *Glottidia pyramidata* (Watabe and Pan, 1984).

from other brachiopods and from several invertebrate phyla. Study of the course of matrix secretion and of $CaCO_3$ deposition will be of interest to establish whether skeletal growth in *Crania* is incremental or nonincremental.

III. Zonation

The inarticulates *Lingula* and *Glottidia* exhibit two types of zonation and two types of markings in their shell structure. One type of zonation is the mineral layer–organic layer–mineral layer alternation also present in molluscan nacreous shell (Chap. 14,VA). The other type is the zonation within a single mineral layer which is expressed differently in *Lingula* and *Glottidia* (Iwata, 1981; Watabe and Pan, 1984). These two types of zonation are evidence of repetitive sequential changes in the activity of the mineralizing cells. At least three types of processes are involved, i.e., the movement of mineral ions to the site of deposition, the formation of protein matrix by mantle cells, and enzyme secretion associated with the development of hydrophobic layers. By assuming either a given sequence of these processes or some fixed relative rates of secretion it is possible to devise various hypotheses to account for shell zonation. Pan and Watabe (1988a) have suggested that in *Glottidia* shell the layering of the secondary layer is caused by the alternate secretion of one epithelial cell type that forms the chitin layers and a second cell type which deposits the material between the chitin layers. Similar alternations in secretory activities have been observed during regeneration of the shell of this animal (Pan, 1985; Pan and Watabe, 1989). In addition to zone formation, the individual mineral units of the primary and secondary layers of the articulate *Notosaria* show transverse markings (Williams, 1971). The markings and changes in mineral density within a single layer suggest a nonuniformity in the rate of ion transport by the mantle epithelial cells. A fourth type of nonuniform mineral deposition occurs as the mantle extends its edge and deposits new shell. The growth in area occurs in increments, as can be seen in the periostracum and outer part of the primary layer of *Glottidia*. Watabe and Pan (1984) have reported that in a 2-cm-long shell, it is possible to show some 1000 distinct lines with numerous minor lines between.

IV. Ion Transport

Calcium and phosphate uptake rates and their distribution among the tissues have been studied in *Glottidia* by using ^{45}Ca and ^{32}P (Pan and

Figure 15.7 (a) Uptake of ⁴⁵Ca by *Glottidia pyramidata*. (b) Uptake of ³²P by *Glottidia pyramidata* (Pan and Watabe, 1988b).

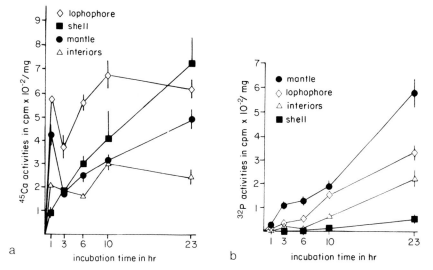

Watabe, 1988b). Figure 15.7a shows a rapid uptake of ⁴⁵Ca by the lophophore (the feeding organ) and the mantle followed by a decrease, which would result from its movement into other organs and the shell. The mantle and shell then continued their uptake. The rate of calcium deposition in the shell was calculated to be 1.2 μg g^{-1} h^{-1}. When ³²P was followed in the same way, the relationships were different (Fig. 15.7b). The rate of phosphorus deposition was approximately 1/300 that of calcium. The reason for the marked difference in shell deposition probably lies in the much greater concentration of phosphorus in tissue compounds and in metabolic reactions.

V. Summary

1. Brachiopods, like molluscs, have an epithelial system of shell formation consisting of (1) a mantle bounded by a single layer of epithelium, (2) a tanned periostracum that covers the outer shell surface, and (3) an extrapallial fluid between the mantle and the periostracum. The epithelial cells transport ions and organic compounds into the extrapallial fluid, and it is from these materials that shell mineral and organic matrix are formed.
2. The shell mineral is layered and commonly of two structural types.

Specialized areas of the mantle epithelium are responsible for each type. The mechanisms that underly the layering and incremental growth of the exoskeleton involve a delicate control of timing, of organic synthesis and secretion, and of ion transport. The details of these mechanisms apply to the entire field of biomineralization and are yet to be explained.

3. The class Articulata deposits shells of calcium carbonate (calcite) whereas the class Inarticulata has shells of calcium fluorapatite. No other multicellular invertebrate phylum has a phosphatic exoskeleton, although deposits of calcium phosphate may be formed.

4. The inner shell layer of Articulata is of particular interest in that it is constructed of long calcite fibers, each of which is said to be initiated and formed by a single epithelial cell. Somewhat similar structures constitute the prismatic layer of molluscan shell but, in contrast to the brachiopod fibers, these are formed from the extrapallial fluid and grow inward from the periostracum.

5. Generalizations concerning the mineral ultrastructure of shells of Articulata and Inarticulata are scarcely warranted on the basis of our current knowledge. However, in the genera discussed, the mineral structure in Articulata is clearly more precisely controlled, suggesting as one possible explanation a more definite ordering of the organic molecules of the matrix surrounding the crystals. Spherulites which constitute an important part of *Lingula* and *Glottidia* shells require little or no orientation of matrix molecules.

6. The organic content of phosphatic shells may be many times that of shells of calcium carbonate and the amino acid composition is different in the two shell types.

7. Experimental studies of brachiopod mineralization are few, as one might expect in view of the paucity and relative inaccessibility of specimens. However, a few species, including those with phosphatic shells, are available for comparative studies of invertebrate phosphatic mineralization, about which so little information is available.

References

Gaspard, D. (1978). Biomineralisation chez les brachiopods articules. Microstructure et formation de la coquille. *Ann. Paleontol.* **64**, 1–25.

Iwata, K. (1981). Ultrastructure and mineralization of the shell of *Lingula unguis* Linne (inarticulate brachiopod). *J. Fac. Sci. Hokkaido Univ.* **20**, 35–65.

Jope, M. (1977). Brachiopod shell proteins: their functions and taxonomic significance. *Am. Zool.* **17**, 133–140.

Jope, M. (1979). The protein of brachiopod shell—6. C-terminal end groups and sodium dodecyl sulfate–polyacrylamide gel electrophoresis: Molecular structure of the protein. *Comp. Biochem. Physiol. B* **63**, 163–173.

LeGeros, R. Z., Pan, C. M., Suga, S., and Watabe, N. (1985). Crystallochemical properties of apatite in atremate brachiopod shells. *Calcif. Tissue. Int.* **37**, 98–100.

McConnell, D. (1963). Inorganic constituents in the shell of the living brachiopod *Lingula*. *Geol. Soc. Am. Bull.* **74**, 363–364.

Moore, R. C. (ed.) (1965). *Treatise Invert. Paleontol.* **1**.

Pan, C.-M. (1985). "Ultrastructural and Physiological Investigations of the Mineralization of the Inarticulate Brachiopod, *Glottidia pyramidata* (Stimpson)," Ph.D. dissertation. Univ. of South Carolina, Columbia, South Carolina.

Pan, C.-M., and Watabe, N. (1988a). Shell growth of a lingulid *Glottidia pyramidata* Stimpson (Brachiopoda, Inarticulata). *J. Exp. Mar. Biol. Ecol.* **118**, 257–268.

Pan, C.-M., and Watabe, N. (1988b). Uptake and transport of shell minerals in the lingulid *Glottidia pyramidata* (Brachiopoda, Inarticulata). *J. Exp. Mar. Biol. Ecol.* **119**, 43–53.

Pan, C.-M., and Watabe, N. (1989). Periostracum formation and shell regeneration in the lingulid *Glottidia pyramidata* (Brachiopoda: Inarticulata). *Trans. Am. Micros. Soc.* In press.

Stricker, S. A., and Reed, C. G. (1985). The ontogeny of shell secretion in *Terebratalia transversa* (Brachiopoda, Articulata). II. Formation of the protegulum and juvenile shell. *J. Morphol.* **183**, 251–271.

Tanaka, K., Ono, T., and Katsura, N. (1988). A hydroxyapatite-adsorbable protein complex in the shell of *Lingula unguis*. *Jap. J. Oral Biol.* **30**, 219–226.

Tuross, N. (1986). The proteins of the shell of the brachiopod, *Lingula*. *In* "Symposium on the Origins of Ocean Chemistry and Its Significance to Biomineralization." Univ. of Texas at Arlington, Arlington, Texas. (Abstr.)

Watabe, N., and Pan, C.-M. (1984). Phosphatic shell formation in atremate brachiopods. *Am. Zool.* **24**, 977–985.

Williams, A. (1965). Introduction. *Treatise Invert. Paleontol.* **1**, H1–H5.

Williams, A. (1970). Spiral growth of the laminar shell of the brachiopod *Crania*. *Calcif. Tissue Res.* **6**, 11–19.

Williams, A. (1971). Comments on the growth of the shell of articulate brachiopods. *Smithson. Contrib. Paleobiol.* **3**, 47–67.

Williams, A. (1977). Differentiation and growth of the brachiopod mantle. *Am. Zool.* **17**, 107–120.

Williams, A., and Rowell, A. J. (1965). Brachiopod anatomy. *Treatise Invert. Paleontol.* **1**, H16–H57.

enclosed sacs and canals. These are the sites of the senses of balance and hearing and as such they lie buried in the ear region of the skull with only an endolymphatic duct to indicate their origin (Fig. 16.1). Part of this sensory system contains large calcareous deposits which stimulate hair cells when the head is moved. Little is known about how these otoliths are formed, but they are surrounded by a particular fluid, the endolymph. This is unusual as an extracellular secretion in that it is high in potassium and low in sodium ion concentrations. It also has a potential difference from 10–100 mV positive in relation to the surrounding perilymph. The walls of the labyrinth contain the enzyme carbonic anhydrase, but whether this is associated with the calcification of the otoliths or is involved in other ion movements is not clear (Simkiss, 1967).

A. Otoliths

Crystallographic analysis of the otoliths provides an interesting series of phylogenetic changes. The agnathans produce otoliths of poorly crystal-

Figure 16.1 Structure of the inner ear of a vertebrate. The maculae are sensory areas that are stimulated when the animal moves by the otoliths that rest upon them. (From Simkiss, 1967).

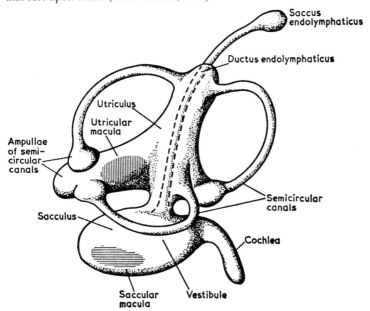

line apatite; those of fish and amphibians are mainly aragonitic; while the reptiles, birds, and mammals have calcitic otoliths (Carlstrom, 1963). Vaterite is found in the otoliths of the lamprey and sturgeon.

Examination of the calcitic otoliths of the rat by very-high-resolution electron microscopy has enabled the investigation of individual crystallites by electron diffraction (Mann et al., 1983). When studied in this way, this calcite appears to be unusual in the number of defects that it has in the lattice and in the features at the edges of the crystals. By comparing the incidence of crystal faces in synthetically derived calcite and aragonite with the same minerals in the otoliths of fish, amphibians, and reptiles, it became apparent that the major planes observed by indexing are not the same (Table 16.1). They concluded that man-made calcium carbonates were probably formed by the reactions:

$$Ca^{2+} + HCO_3^- \rightarrow H^+ + [CaCO_3] \text{ (hydrated ion pairs)} \tag{16.1}$$

$$[CaCO_3] + [CaCO_3] + Ca^{2+} + CO_3^{2-} \rightarrow CaCO_3 \cdot 6H_2O \text{ (amorphous)} \tag{16.2}$$

$$CaCO_3 \cdot 6H_2O \rightarrow \text{crystalline } CaCO_3 \tag{16.3}$$

Biogenic carbonates may be seeded in this way but the development of particular lattice faces suggests that in otoliths there is probably control by the association of calcium and/or carbonate with an organic polymer.

The biological influences on otolith growth are of considerable interest, particularly in fish where they are widely used to age individuals. In fish from temperate waters there is a seasonal variation in the rate of otolith growth and the deposition of matrix. As a result there are annual rings that can often be enhanced by burning, so that the thickness can be measured to give an indication of relative growth. These bands are used to calculate the productivity of commercial fish stocks (Ricker, 1968). It

Table 16.1 Incidence of the Main Crystal Faces in Normal Aragonite and Calcite Compared with those in the Otoliths of Varius Species[a]

Crystal Face	Aragonite	Calcite	Fish	Frog	Rat
001	—	*	*	*	*
010	*	—	*	*	—
011	—	—	—	*	—
111	—	—	—	—	*
211	—	*	—	—	—
421	—	*	—	—	*

[a] After Mann et al. (1983).

is, however, difficult to devise fishery programs for fish which migrate to different oceans and grow at various rates. It has, therefore, been suggested that the types of crystal and the traces of other elements incorporated into the otolith during its growth may yet indicate at what temperature and in what ocean the fish grew before they were caught (Gauldie et al., 1980; Gauldie, 1986). Support for studies on the finer details of such otolith growth was initiated by the finding that there are daily increments on the otoliths, so that the careful study of these may reveal a whole range of physiological incidents that have occurred to the fish and affected its growth rate (Panella, 1971).

B. Otoconia

In many of the lower vertebrates, but especially in the amphibia, the endolymphatic system of calcium carbonate deposition becomes enormously extended. The endolymphatic duct of the larval tadpole passes through a foramen in the skull and comes to lie over the brain. It con-

Figure 16.2 Transverse section of a metamorphosing clawed toad (*Xenopus laevis*) showing how the endolymphatic sac extends from the inner ear and comes to lie over the brain. (From Simkiss, 1967.)

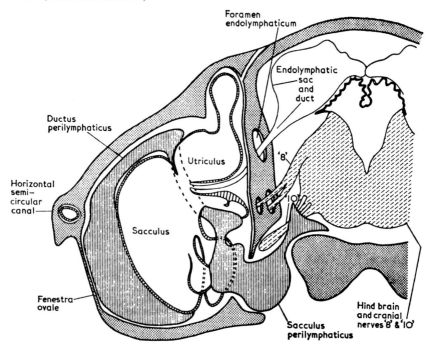

tinues to expand and passes backward over the spinal cord and may, in the adult, emerge between the vertebrae. This enormously extended organ retains the ability to secrete small granules of calcium carbonate or otoconia so that, during the larval phase, the animal stores large quantities of aragonite in this form. This provides a readily available source of calcium during metamorphosis, when the tadpole is unable to feed but is actively forming its adult skeleton. During this phase there is no increase in the calcium content of the tadpole but there is a major shift from the store of aragonitic calcium in the inner ear to the phosphatic calcium of the developing skeleton (Fig. 16.3). A similar movement of calcium from the stores in the endolymphatic system occurs in some reptiles during the calcium stress of egg laying and some snake embryos also store calcium at this site (Simkiss, 1967).

There is, therefore, clear evidence for the physiological control of various minerals in these vertebrates. In the adult frog, the endolymphatic deposits of aragonite can also be used in acid–base balance (Simkiss, 1968a), as a calcium store, and as a source of calcium for bone repair (Schlumberger and Burk, 1953). In an extended series of experiments, Robertson (1970, 1976) has investigated the role of parathyroid hormone and calcitonin in the regulation of these epithelial activities and in the endocrine control of mineral deposition and resorption. In

Figure 16.3   Changes in the quantity of calcium in the hard parts of the tadpole during metamorphosis. In the early phase calcium is stored in the endolymphatic sacs. This is followed by a phase of bone formation but, at metamorphic climax (day +2), the tadpole ceases feeding and further bone formation occurs at the expense of calcium resorbed from the calcareous deposits in the inner ear. (From Simkiss, 1967.)

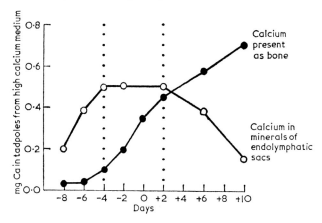

most vertebrates, parathyroid hormone is involved in raising blood calcium levels while calcitonin lowers them and protects the skeleton. There are, however, considerable differences between the responses of different organs in different classes of vertebrates (Dacke, 1979).

II. Eggshell Formation

A second vertebrate organ that uses an epithelial layer as a basis for biomineralization is the oviduct of the reptiles and birds. After ovulation, the avian oocyte passes rapidly down the oviduct and has albumen and shell membranes secreted around it. It is then held at the distal end of the oviduct for periods of 15 to 30 hr, depending on the species, while a calcareous eggshell is formed on the surface of the shell membranes. The phenomenon has been extensively studied and reviewed (Simkiss, 1961, 1967, 1968b; Hamilton, 1986) and we will only comment on the general biology of the process.

Eggshell formation in birds is an extremely rapid process so that in the domestic fowl about 5 g of calcium carbonate is deposited within 20 hr. This is a rate of calcium deposition sufficient to deplete the calcium from the total blood volume within 15 min were it not replenished from dietary and bone supplies. During this process, there is also a change in the acid–base balance of the bird and much of the CO_3^{2-} appears to be formed by the hydration of metabolic carbon dioxide (Lorcher and Hodges, 1969). There is an abundant supply of carbonic anhydrase in the epithelial cells of the shell gland (Gay and Mueller, 1973) which could catalyze the reaction

$$Ca^{2+} + CO_2 + H_2O \rightarrow CaCO_3 + H_2^+$$

The protons that are released during eggshell formation induce an acidosis in the bird. As a result, laying hens often increase their respiratory activity by panting while protons are also excreted by the kidney. The urine of a laying bird goes distinctly acid during the process of eggshell calcification (Simkiss, 1970). The shell gland of the bird is obviously transporting ions across the epithelium to deposit calcium and carbonate ions into the oviduct and also moving protons and counterions outward into the blood. The epithelium develops a small potential difference of about 10 mV (mucosa negative) during this process (Hurwitz et al., 1970).

The calcite crystals that form the avian eggshell develop from a number of discrete nucleation sites on the fibrous shell membranes. These form the centers for spherulitic growth. The sites of this activity are clearly seen in a section of the eggshell and they have long been known as mammillary cores (Fig. 16.4). The mammillary cores are small (10 μm) discrete spheres of protein. In decalcified eggshells they can be seen to be pitted with holes through which the emerging crystals of spherulitic calcite have passed. They appear, therefore, to be one of the most useful sources of material for experimental studies of a protein–crystal interaction. To date, however, they have only been analyzed chemically (Cooke and Balch, 1970) and shown to contain carbonic anhydrase (Krampitz, 1982).

The shell has been studied in considerable detail both crystallographically and in terms of its soluble and insoluble protein matrix (Baker and Balch, 1962; Leach, 1982). Two sets of observations are particularly interesting in the context of biomineralization. The first relates to the distribution of other cations throughout the shell. Magnesium in particular is concentrated in the inner part of the spherulite structure, which strongly suggests that the initial seeds of the shell have a somewhat different

Figure 16.4 Stylized radial section of an avian eggshell showing the structure and composition of the various regions. Organic matrix is shown on the left and crystalline structure on the right. (From Simkiss, 1968b.)

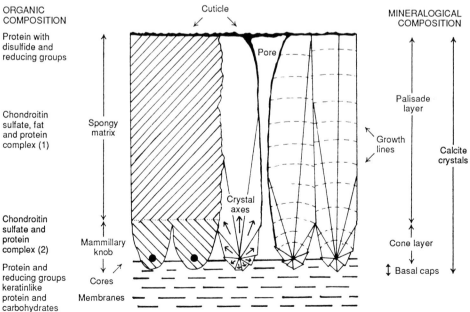

composition or crystal size from the main calcite of the eggshell. The second set of observations is both novel and exciting. It will be apparent that, when a spherulite grows to form an eggshell, the crystallite axes will initially be in all directions. Growth inward is, however, clearly inhibited because the ions are being supplied from the oviduct wall on the outside of the egg. Growth laterally also becomes inhibited as the crystals from the various nucleation sites abut onto each other (Fig. 16.4). As the shell grows outward there is, therefore, an increasingly uniform orientation of the crystals (Fig. 16.5). This can be measured using X-ray diffraction to compare the intensity of the peaks associated with particular crystal planes (I) with those obtained from a sample of ground and randomly oriented crystals (I_o). Using this approach Silyn-Roberts and Sharp (1986) showed that, for a plane at distance x from the site of nucleation and with a radius for a shell column of r, there should be a relationship of $I/I_o = 2x/r$ (Fig. 16.6). Using data obtained from a number of turtle eggshells, they found that the theoretical and experimental results corresponded closely. This shows that the eggshell consists of spherulite crystals that grow in an unhindered way in the 001 direction. Starting from single nucleation sites, these crystals initially extend in all directions but are gradually restricted to growing only in the outward direction.

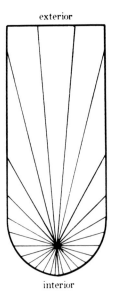

Figure 16.5 The configuration of crystal growth in a shell column based on radial growth from a single nucleation site. (From Silyn-Roberts and Sharp, 1986.)

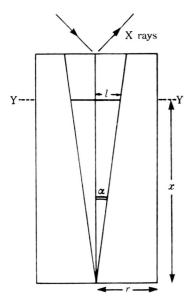

Figure 16.6 Diagram to illustrate the relation, at level Y—Y, between the intensity ratio of the basal plane at distance x from nucleation and radius of the shell column. Crystals in the 001 reflection condition are contained within the cone of semiapical angle α with radius l. (From Silyn-Roberts and Sharp, 1986.)

When these approaches are applied to some avian eggshells, this simple relationship is lost (Silyn-Roberts and Sharp, 1986). In the ostrich (Fig. 16.7) it is clear that at a distance of about 700 to 800 μm from the interior of the shell the initial unhindered crystal growth is suddenly

Figure 16.7 Relation between $\sqrt{(I/I_0)}$ of the 006 plane and the distance x from the interior of the shell of the ostrich. Points 1 and 2 mark the two successive interruptions to 001 texture development. (From Silyn-Roberts and Sharp, 1986.)

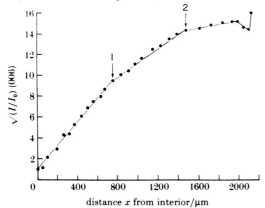

changed. A similar modification occurs at a distance of about 1400 μm. Investigation of the protein distribution in these eggshells showed that the first crystal transition was associated with the presence of fibrous organic sheets while the second correlated with an increase in their density. Turtle eggshells contain an organic matrix but this clearly does not disrupt the crystalline structure. By contrast, it appears that in avian eggshells the matrix modifies the crystal structure, and this correlates with an increase in the strength of the shell. The insoluble matrix appears to act as a reinforcing fibrous network that is deposited in the shell to tie the crystal columns together structurally and impart additional strength. This is an interesting conclusion and this mechanical function clearly needs to be integrated into the concepts of the crystallographic influence of matrix proteins in a variety of other shells.

III. Bone Formation

Bone is a highly organized tissue characterized by an extensive extracellular matrix in which the hydroxyapatite bone mineral [$Ca_{10}(PO_4)_6(OH)_2$] is deposited. There are three main types of cells associated with the production and resorption of this mineral. Osteoblasts are involved in forming the organic matrix and converting it into a mineralized form. Osteocytes are derived from osteoblasts as they become embedded in the mineralized bone. They retain contact with each other via cellular processes and are largely involved in maintaining the bone mineral and controlling its exchange with the plasma. Osteoclasts are large multinuclear cells that occur on the bone surface, often in lacunae that are formed by their activity. They have a ruffled surface that is a site of bone resorption. This is probably achieved by secreting acids, chelating agents, and extracellular proteases onto the bone surface.

   Bone may form in one of two basic ways. In intramembranous ossification the intercellular connective tissue increases in amount and density. Some of the connective tissue cells differentiate to form osteoblasts and shortly afterward bone mineral is deposited. The second type of bone formation is more complex for it is preceded by the formation of a cartilaginous structure. Cartilage cells or chondrocytes divide to form columns between which mineral is deposited. This mineral is, however, resorbed when blood vessels penetrate this region, osteoblasts are formed, and true bone minerals are deposited in the surface layers of the cartilage.

   In most vertebrates the skeleton represents a massive reservoir of calcium, phosphate, and other ions which are in a state of continual interchange with the components of the body fluids. Three processes are involved: (1) accretion or the formation of new bone by osteoblasts, (2) resorption or the destruction of bone by osteoclasts with a return of its ions to the blood, and (3) exchange or the movement of ions between the body fluids and bone crystals with little change in mass. The exchange process is not controlled by any cell type but it is thought that osteocytes facilitate the process by providing channels in the bone and they may also assist in the "osteolysis" of the mineral in a rather poorly defined way.

Among the fish, the Chondrichthyes (sharks, rays, etc.) have a skeleton that consists almost entirely of calcified cartilage. Growth occurs by the deposition of additional cells and mineralized layers but there is no resorption. If, however, ^{45}Ca is injected into a shark it rapidly labels virtually the whole skeleton, implying that most of the mineral is exchangeable and in equilibrium with the blood (Urist, 1961). This is in contrast with the bones of most vertebrates, where only about 2% of the mineral is exchangeable.

In the Osteichthyes or bony fish most of the "advanced" forms have unusual skeletons. Unlike those of most other vertebrates, they are acellular. As the osteoblasts deposit the bone, they move away from the site of mineralization, so that they do not become trapped as osteocytes in the way that they do in the bone of all other vertebrates. This acellular bone appears to have been derived from the more usual cellular bone which still persists in some primitive fish, such as the clupeiforms (e.g., herring). It is not clear why most modern fish have abandoned the ancestral and more normal cellular bone. According to Moss (1962), who studied the repair of fractured bones in fed and starved fish, the cellular bone recovered faster. He considered acellular bone to be "dead" bone in that it could play little part in the normal physiology of the animal. In contrast, Brehe and Fleming (1976) found that 80% of the calcium in the acellular bone of the killifish (*Fundulus kansae*) was diffusible and could still be mobilized according to the needs of the animal.

If acellular bone is difficult to understand physiologically, it is useful in drawing attention to two problems that warrant further consideration. First, bone remodeling is usually considered as a way of maintaining a supply of ions that facilitate the growth and repair of the skeleton. Fish, however, may have adequate access to these ions from water and dietary intakes (Simkiss, 1974). The absorptive capacity of the gills and intestine may be more than adequate to supply these ions to the blood at concentrations sufficient to maintain bone formation. Second, the concept of a layer of osteoblasts moving away from the mineralization front is an interesting one. In most vertebrate skeletons the new layers of tissue are formed by a continuous layer of cells which isolates a separate compartment in which matrix is deposited and mineralized. In some systems, such as tooth enamel, the primary purpose of these layers appears to be to regulate the availability of lattice ions. If the integrity of these cellular layers is disrupted, mineralization occurs rapidly and in an unorganized way. In fact, the only tissue where the cells are separated and do not form an anatomical compartment in which they deposit matrix and mineral is cartilage and this is associated with a temporal sequence of mineralization changes (Simkiss, 1982). Acellular bone is, perhaps, the clearest way of picturing a cellular layer which delimits a compartment within which the mineralization process proceeds.

A. Bone Mineral

The mineral hydroxyapatite, with a unit cell of $Ca_{10}(PO_4)_6(OH)_2$, is the basic material of bones and teeth. It is often somewhat modified with fluoride or chloride ions substituting for hydroxyl groups or strontium replacing calcium in isomorphic substitutions. Carbonate may also replace the tetrahedral PO_4 group. In its pure state, $Ca_{10}(PO_4)_6(OH)_2$ is a monoclinic structure with a Ca : P molar ratio of 1.66 (10/6), but in most of the substituted forms, the crystal is hexagonal and the molar ratios often change considerably. If hydroxyapatite is precipitated from a highly supersaturated solution at 37°C an unstable amorphous mineral is formed with a Ca : P ratio of 1.5. This slowly converts to hydroxyapatite. These minerals occur as crystals about $20 \times 5 \times 5$ nm in size, i.e., they are sufficiently small to be in the range where solubility depends on crystal size. This makes it difficult to determine a solubility product constant for bone, and when freshly precipitated hydroxyapatite is kept in solution the crystals actually increase in size and modify their stoichiometry. This is one of the effects of adding fluoride to drinking water. The fluoride ion substitutes for hydroxyl groups in bone mineral, and this leads to a less soluble mineral partly because of the change in crystal lattice and partly because it produces larger crystals.

The formation of bone mineral with various Ca : P ratios, the variety of their solubility products, and the possibility that there may be a precursor mineral phase have all been complicating influences in the understanding of bone formation. The earliest mineral that can be detected in embryonic chick bone is noncrystalline, suggesting an amorphous phase, while the first material to give X-ray diffraction appears to be brushite ($CaHPO_4\cdot2H_2O$). There has, however, been considerable argument as to whether or not this is a preparative artifact. Similar questions have been raised regarding the reported presence of octacalcium phosphate [$Ca_8H_2(PO_4)_6\cdot5H_2O$) and these issues are still controversial. The problems are important, however, because of the light they may throw on the biomineralization process. If an amorphous precursor is formed, it implies that bone forms in a region that is very highly supersaturated. If octacalcium phosphate is the first mineral to form, the solution is probably slightly acid whereas brushite would imply a more acid fluid. If hydroxyapatite formed directly, then only a low degree of supersaturation would be envisaged (Posner, 1985). The implications of these minerals in terms of theories of biomineralization involving compartment sizes, effects of ion pumps, and types of nucleating surfaces are enormous.

B. Bone Matrix

On a dry weight basis, bone is roughly 60–70% mineral embedded in an organic matrix consisting largely of collagen (Table 16.2). As this matrix has been analyzed in more and more detail an ever-increasing list of constituents has been revealed, almost all of which provide the basis for some theory of bone formation.

Collagen is a highly characteristic group of fibrous proteins found in all multicellular animals. They account for about 25% of all the protein

Table 16.2 Relative Proportions of Bone Matrix Components[a]

Constituent	Percentage by Weight
Collagen	88.0
Osteonectin	2.5
α-Carboxyglutamic acid-containing protein	1.5
Sialoprotein	1.0
Structural glycoprotein	1.0
Proteoglycans	1.0
Peptides	0.8
Lipids	0.4
α_2-HS-glycoprotein	0.4
Albumin	0.3
Proteolipids	0.3
Phosphoprotein	0.2
Bone morphometric protein	0.1

[a] After Urist *et al.* (1983).

present in a mammal and have an intimate relationship with the mineralized skeleton. It must be stressed, however, that not all collagens are
capable of becoming calcified. The characteristic feature of collagens is
the α chain, which is wound in a triple helix to form a small, ropelike
structure roughly 300 nm by 1.5 nm (Fig. 16.8). There are seven distinct
types of α chain, so that over a hundred different combinations would
be possible. Of this wide variety, relatively few are found and about 90%
of the collagen in the body is of the so-called type I, consisting of two α_1
chains and one α_2 chain in each molecule.

The collagen molecule has two unusual features. First, the α chain is
so tightly coiled that every third amino acid is glycine, i.e., the simplest
amino acid and the only one without a side chain that could fit into this
triple helix. Second, collagen contains hydroxyproline and hydroxylysine, that are rarely found in other proteins. These hydroxy groups help
to stabilize the triple helix. In addition, once the collagen fibers have
been secreted from a cell into the extracellular space they become
strengthened by cross-linking. Some of the lysine and hydroxylysine
residues are deaminated in this process to yield highly reactive aldehyde
groups which form covalent bonds between the molecules. The significance of this is apparent when the collagen fibril is studied. This fibril
has a 67-nm pattern that arises from a "stagger" in the assembly of
adjacent collagen molecules (Fig. 16.8). A 67-nm stagger to an array of
molecules each of which is 300 nm long means, however, that after four
spacings ($67 \times 4 = 268$) there is only 32 nm of molecule left to fit the next
67-nm stagger. There are, therefore, 35 nm gaps between the heads and
tails of the collagen molecules. The collagen fiber is therefore held together by cross-links and each of the individual molecules is separated
from the molecules in front and behind it by a small space. This model of
the collagen molecule produced great excitement when it was first published by Hodge and Petruska (1963). Electron micrographs of mineraliz-

Figure 16.8 The arrangement of tropocollagen molecules in a collagen fiber showing the hole between the staggered units and its
possible involvement in bone mineral formation.

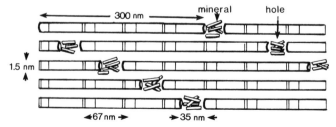

ing connective tissue had shown that bone crystallites occurred at regular intervals along the collagen fiber, and it was suggested that the gaps might be the sites of mineral deposition in bone. Neutron and X-ray scattering experiments have, in fact, confirmed that the calcium-hydroxyapatite crystals do occur in the spaces between collagen units and the mineral does appear to be nucleated at a specific, though unknown, location within the head-to-tail gap (Miller, 1984). There are, however, problems. Not all type I collagens calcify and the mineral that does form is not always in these gaps. It is possible, therefore, that intermediary macromolecules attach to the collagen and convert it into a mineralizing system. Most of the noncollagenous molecules listed in Table 16.2 have been studied with this possibility in mind.

One protein that has attracted extensive attention is osteocalcin. This bone-specific protein contains three residues of α-carboxyglutamic acid (GLA). It is formed by vitamin K-dependent carboxylation of glutamic acid, which bestows strong Ca-binding properties on the molecule that could therefore induce mineralization (Gallop et al., 1980). A similar phenomenon has been attributed to acidic phospholipids such as phosphatidylserine, which forms calcium–acid phospholipid–phosphate complexes (Yaari et al., 1984). It has also been suggested that phosphorylation of serine and threonine residues by the osteoblast cells may produce phosphoproteins which bind to particular sites at the hole zones of the collagen fibrils (Glimcher, 1984). All of these approaches attribute the onset of calcification to the presence of calcium binding sites in appropriate regions of the collagen molecule. A somewhat different set of properties was revealed with the analysis of a 32-KDa protein, osteonectin. This protein not only binds to collagen and renders it capable of inducing the nucleation of bone mineral but it also binds directly to hydroxyapatite crystals (Termine et al., 1981). There are, therefore, many components of the collagen-associated matrix that are capable of facilitating calcification and DNA sequencing studies are beginning to show how these may have originated (Bolander et al., 1988). Much is made in the literature about their specificity either in inducing mineral deposition or in occurring only at particular sites. Despite this, most physiologists also invoke some form of inhibition as a way of controlling the mineralization process.

C. Inhibitors of Bone Formation

The problem of attributing complete specificity in the initiation of bone formation to specific sites in the organic matrix has led to the suggestion that there may be a variety of inhibitory molecules that can be selectively removed. The first of these to be identified was pyrophosphate (Fleisch and Neuman, 1961). These ions occur in plasma, urine, and saliva at concentrations of 10^{-6} mol dm^{-3} which are capable of inhibiting most forms of calcification. Pyrophosphate is hydrolyzed to orthophosphate by a variety of enzymes, such as alkaline phosphatase, but since these enzymes also occur at sites which do not calcify it is suggested that other inhibitors may also be present. Ions such as citrate and magnesium, proteins such as albumin, and polyelectrolytes such as glycosaminoglycans have all been implicated in such functions. Some of these inhibitors may be components of the matrix, produced and acting locally (Russell *et al.*, 1986).

D. Cellular Systems in Bone Formation

It is generally recognized that the osteoblast synthesizes type I collagen. It probably also synthesizes glycoproteins, osteocalcin, osteonectin, phosphoproteins, and many of the other components of the matrix. In culture, osteoblasts produce molecules that regulate bone growth and one, bone morphogenic protein, is capable of inducing active bone formation when introduced into connective tissue (Urist *et al.*, 1983). In intact bone there is often an interaction whereby osteoblasts that are stimulating bone formation at one site appear to induce osteoclasts to cause bone resorption at an adjacent site. Indeed, Howard *et al.* (1981) claim to have characterized such a coupling factor which coordinates the activities of these two cell types. Clearly, as with all forms of biomineralization, cells play a dominant regulatory role. There have, however, also been a number of hypotheses which further accentuate this cellular involvement. The first of these was Howard's (1956) bone membrane concept, initially proposed because, as a clinician, he could find little correlation between the composition of the blood and the solubility of bone minerals. The idea was supported by Neuman and Ramp (1971), who found that bone potassium was not in equilibrium with serum potassium, but the theory was stated most clearly by Talmage (1969). It has five postulates (Fig. 16.9). There are

1. Bone contains a fluid compartment separated from the blood by a layer of cells.
2. The mineralized tissue is associated with the inner fluid which has a calcium and phosphate content controlled by the solubility of the bone mineral.
3. The outer fluid compartment is in equilibrium with the blood.

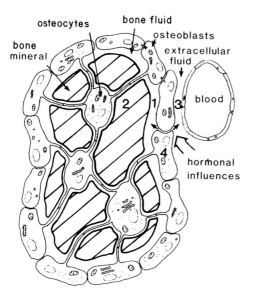

osteocytes   bone fluid
              osteoblasts
bone   extracellular
mineral   fluid
              blood
              hormonal
              influences

Figure 16.9 The bone–membrane concept in which newly forming bone is enclosed by a cellular layer. As a result, there is an inner fluid compartment (1) in equilibrium with the bone mineral (2) and separated from the body fluids (3) by a sheet of bone cells (4).

4. The movement of calcium and phosphate ions between these two fluids will be controlled by ion pumps in the bone membrane.
5. These cellular activities will be regulated by parathyroid and calcitonin hormones.

The similarity between this hypothesis and those of most other forms of biomineralization is very clear since they all emphasize an isolated compartment, solubility products, a matrix, and ion pumps. The bone membrane concept is not, however, very popular largely because of the problems of identifying the system anatomically. Instead, most cellular theories of bone formation rely on diffusion barriers as an alternative to a living membrane.

Such a system is invoked in the matrix vesicle theory. Matrix vesicles were first described in epiphyseal cartilage but have since been found in lamellar bone, dentine, antlers, and at a variety of pathological sites (Anderson, 1985). They are small round organelles of about 100 nm in diameter that are thought to be derived from cell processes originating at the osteoblast plasma membrane. Several enzymes, including alkaline phosphatase and nucleoside triphosphate (NTP) pyrophosphatase, are associated with these vesicles, which are found in the matrix at sites of mineralization. It has been suggested that the alkaline phosphatase in these vesicles may increase the supply of phosphate ions, transport calcium into the vesicle, or remove inhibitors such as pyrophosphate ions from sites of calcification. Alternatively, NTP pyrophosphatase, which catalyzes the reaction

$$ATP \rightarrow AMP + P_2O_7$$

may initiate the precipitation of amorphous calcium phosphate via the hydrolysis of pyrophosphate (Hsu and Anderson, 1984). By various schemes, therefore, it has been suggested that enzymes in these cellular bodies initiate the formation of bone mineral at sites adjacent to the matrix vesicle and this is in keeping with what is observed by electron microscopy. Matrix vesicles are found to contain calcium and phosphate, often in the form of crystals that pierce the membrane and extend into the matrix. Perhaps one of the most interesting interpretations of this phenomenon, which puts it immediately into the area of general cell biology, is the apoptosis concept of Kardos (1986). A large number of cellular activities involve changes in the distribution of cytoplasmic calcium. In response to a rapid accumulation of calcium, many cells have been found to redistribute the cytoplasm and discard a portion of it in which calcium is concentrated. This cell deletion or apoptosis process occurs during the development and maintenance of many tissues and it is suggested that matrix vesicles have much in common with this cellular response which subsequently triggers bone formation.

Bone, then, represents an unusual form of biomineralization. Traditionally, its formation has been regarded as a largely extracellular phenomenon and the search for ion pumps and cellular compartments has tended to be secondary to studies on mineral composition, matrix components, and endocrinology. Bone physiology is undoubtedly the source of many of the concepts of biomineralization, but it remains as one of the few physiological processes for which there is no unified theory and one of the few medical interests that has not benefited from comparative studies.

IV. Summary

1. The vertebrates produce a variety of biomineralized products, including the calcareous deposits in the inner ear and the shells around the eggs of various reptiles and birds. Vertebrate skeletons are always extracellular and are composed of calcium phosphates in the form of hydroxyapatites.

2. Otoliths are of interest as the products of epithelial secretion in which there is a clear influence of the matrix on mineral shape.

3. Eggshells are also secreted by an epithelial system but the obvious nucleation sites and influence of matrix proteins on spherulitic crystal growth make them ideal objects for crystallographic studies.

4. Skeletal tissues range from calcified cartilage and acellular bone to dentine, enamel, and various types of cellular bone.

5. Bone mineral consists of hydroxyapatite which, in mature bone, consists of $Ca_{10}(PO_4)_6(OH)_2$. There is discussion as to whether there is a precursor mineral which transforms into bone mineral during the biomineralization process.

6. Bone matrix is largely composed of collagen fibers. One of the initial sites of calcification appears to be the gap between "head and tail" collagen molecules. Since many tissues also contain type I collagen which does not calcify, it is likely that one or more of the noncollagenous proteins modulates the collagen of bone and assists in the induction of mineralization.

7. A number of small molecules act as inhibitors to bone formation and are presumably removed from sites of ossification.

8. The basic role of the major bone cells, i.e., osteoblasts, osteocytes, and osteoclasts, is clearly understood but cellular influences in the formation of extracellular mineral are still problematical.

References

Anderson, H. C. (1985). Matrix vesicle calcification; a review and update. *Bone Miner. Res.* **3**, 109–149.

Baker, J. R., and Balch, D. A. (1962). A study of the organic material of hens' eggshell. *Biochem. J.* **82**, 352–361.

Bolander, M. E., Young, M. F., Fisher, L. W. Yamada, Y. and Termine, J. D. (1988). Osteonectin cDNA sequence reveals potential binding regions for calcium and hydroxyapatite and shows homologies with both a basement membrane protein (SPARC) and a serine protease inhibitor (ovomucoid). *Proc. Nat. Acad. Sci. U.S.A.,* **85**, 2919–2923

Brehe, J. E., and Fleming, W. R. (1976). Calcium moblization from acellular bone and effects of hypophysectomy on calcium metabolism of *Fundulus kansea. J. Comp. Physiol.* **110**, 159–169.

Carlstrom, D. (1963). A crystallographic study of vertebrate otoliths. *Biol. Bull. (Woods Hole, Mass.)* **125**, 441–463.

Cooke, A. S., and Balch, D. A. (1970). Studies of membrane, mamillary core and cuticle of the hen eggshell. *Br. Poult. Sci.* **11**, 345–352.

Dacke, C. G. (1979). "Calcium Regulation in Sub-mammalian Vertebrates." Academic Press, New York.

Fleisch, H., and Neuman, W. F. (1961). Mechanisms of calcification: Role of collagen, polyphosphates and phosphatase. *Am. J. Physiol.* **200**, 1296–1300.

Gallop, P. M., Lian, J. B., and Haushka, P. V. (1980). Carboxylated calcium-binding proteins and vitamin K. *N. Engl. J. Med.* **302**, 1460–1466

Gauldie, R. W. (1986). Vaterite otoliths from chinook salmon (*Oncorhynchus tshawytscha*). *N.Z. J. Mar. Freshwater Res.*, **20**, 209–217.

Gauldie, R. W., Graynoth, E. J., and Illingworth, J. (1980). The relationship of the iron content of some otoliths to temperature. *Comp. Biochem. Physiol. A* **66**, 19–24.

Gay, C. V., and Mueller, W. J. (1973). Cellular localization of carbonic anhydrase in avian tissues by labelled inhibitor autoradiography. *J. Histochem. Cytochem.* **21**, 693–702.

Glimcher, M. (1984). Recent studies of the mineral phase in bone and its possible linkage to the organic matrix by protein-bound phosphate bonds. *Philos. Trans. R. Soc. London, Ser. B.* **304**, 479–506.

Hall, B. K. (1978). "Development and Cellular Skeletal Biology." Academic Press, London.

Halstead, L. B. (1987). Evolutionary aspects of neural crest-derived skeletogenic cells in the earliest vertebrates. *In* "The Neural Crest" (P. Maderson, ed.), pp. 339–358. Wiley, New York.

Hamilton, R. M. G. (1986). The microstructure of the hens' eggshell. A short review. *Food Microstruct.* **5**, 99–100.

Hodge, A. J. and Petruska, J. (1963). Recent studies with the electron microscope on ordered aggregates of the tropocollagen macromolecules. *In* "Aspects of Protein Structure" (G. A. Ramachandran, ed.), pp. 289–300. Academic Press, London.

Howard, G. A., Bottemiller, B. L., Turner, R. T., Rader, J. I., and Baylink, D. J. (1981). Parathyroid hormone stimulates bone formation and resorption in organ culture: Evidence for a coupling mechanism. *Proc. Natl. Acad. Sci. U.S.A.* **38**, 3204–3208.

Howard, J. E. (1956). Present knowledge of parathyroid function with especial emphasis upon its limitations. *In* "Bone Structure and Metabolism" (G. E. W. Wolstenholme and C. M. O'Connor, eds.), pp. 206–217. Churchill, London.

Hsu, H. T., and Anderson, H. C. (1984). The deposition of calcium pyrophosphate and phosphate by matrix vesicles isolated from fetal bovine epiphyseal cartilage. *Calcif. Tissue Int.* **36**, 615–621.

Hurwitz, S., Cohen, I., and Bar, A. (1970). The transmembrane electrical potential difference in the uterus (shell gland) of birds. *Comp. Biochem. Physiol.* **35**, 873–878.

Kardos, T. B. (1986). Initiation of mineralization. *N.Z. Dent. J.* **82**, 71–74.

Krampitz, G. P. (1982). Structure of the organic matrix in mollusc shells and avian eggshells. *In* "Biological Mineralization and Demineralization" (G. H. Nancollas, ed.), pp. 219–232. Springer-Verlag, Berlin.

Leach, R. M. (1982). Biochemistry of the organic matrix of the eggshell. *Poult. Sci.* **61**, 2040–2047.

Lorcher, K., and Hodges, R. D. (1969). Some possible mechanisms of formation of the carbonate fraction of eggshell calcium carbonate. *Comp. Biochem. Physiol.* **28**, 119–128.

Mann, S., Parker, S. B., Perry, C. C., Ross, M. D., Skarnulis, A. J., and Williams, R. J. P. (1983). Problems in the understanding of biominerals. *In* "Biomineralization and Biological Metal Accumulation" (P. Westbroek and E. W. de Jong, eds.), pp. 171–183. Riedel, Dordrecht, The Netherlands.

Miller, A. (1984). Collagen: The organic matrix of bone. *Philos. Trans. R. Soc. London, Ser. B.* **304**, 455–477.

Moss, M. L. (1962). Studies of the acellular bone of teleost fish. II. Response to fracture under normal and acalcemic conditions. *Acta Anat.* **48**, 46–60.

Neuman, W. F., and Ramp, W. K. (1971). The concept of bone membrane. Some implications. In "Cellular Mechanisms of Calcium Transfer and Homeostasis" (G. Nichols and R. H. Wasserman, eds.). Academic Press, London. 197–206.

Panella, G. (1971). Fish otoliths: Daily growth layers and periodical patterns. Science 173, 1124–1126.

Posner, A. S. (1985). The mineral of bone. Clin. Orthop. 200, 87–99.

Ricker, W. E. (ed.) (1968). Methods for assessment of fish production in freshwaters. In "International Biological Programme Handbook 3." Blackwell, Oxford, England.

Robertson, D. R. (1970). The ultimobranchial body in Rana pipiens. XI. Response to increased dietary calcium—Evidence of possible physiological function. Endocrinology, (Baltimore) 87, 1041–1050.

Robertson, D. R. (1976). Diurnal and lunar periodicity of intestinal calcium transport and plasma calcium in the frog Rana pipiens. Comp. Biochem. Physiol. A 54, 225–231.

Russell, R. G. G., Caswell, A. M., Hearn, P. R., and Sharrard, R. M., (1986). Calcium in mineralized tissues and pathological calcification. Br. Med. Bull. 42, 435–446.

Schlumberger, H. G., and Burk, D. H. (1953). Comparative study of the reactions to injury. 2. Hypervitaminosis D in the frog with special reference to the lime sacs. Arch. Pathol. 56, 103–124.

Silyn-Roberts, H., and Sharp, R. M. (1986). Crystal growth and the role of the organic network in eggshells. Proc. R. Soc. London, Ser. B. 227, 303–324.

Simkiss, K. (1961). Calcium metabolism and avian reproduction. Biol. Rev. 36, 321–367.

Simkiss, K. (1967). "Calcium in Reproductive Physiology," pp. 1–265. Chapman & Hall, London.

Simkiss, K. (1968a). Calcium and carbonate metabolism in the frog (Rana temporaria) during respiratory acidosis. Amer. J. Physiol. 214, 627–633.

Simkiss, K. (1968b). The structure and formation of the shell and shell membranes. In "Egg Quality. A Study of the Hen's Egg" (T. C. Carter, ed.), pp. 3–25. Oliver & Boyd, Edinburgh, Scotland.

Simkiss, K. (1970). Sex differences in the acid–base balance of adult and immature fowl. Comp. Biochem. Physiol. 34, 777–788.

Simkiss, K. (1974). Calcium metabolism of fish in relation to ageing. In "The Ageing of Fish" (T. Bagenel, ed.), pp. 1–12. Butterworths, London.

Simkiss, K. (1982). Mechanisms of mineralization (normal). In "Biological Mineralization and Demineralization" (G. H. Nancollas, ed.), p. 351–363. Springer-Verlag, Berlin.

Talmage, R. V. (1969). Calcium homeostasis–calcium transport–parathyroid action. Clin. Orthop. 67, 210–224.

Termine, J. D., Kleinman, H. K., Whitson, S. W., Conn, K. M., McGarvey, M. L., and Martin, G. R. (1981). Osteonectin, a bone specific protein linking mineral to collagen. Cell 26, 99–105.

Urist, M. R. (1961). Calcium and phosphorus in the blood and skeleton of the elasmobranchii. Endocrinology (Baltimore) 69, 778–801.

Urist, M. R., Delange, R. J., and Finerman, A. M. (1983). Bone cell differentiation and growth factors. Science 220, 680–682.

Vaughan, J. (1981). "The physiology of bone," 3rd Ed. Oxford Sci. Pub., Oxford, England.

Yaari, A. M., Bosken, A. L., and Shapiro, I. M. (1984). Phosphate modulation of calcium transport by a calcium–phospholipid–phosphate complex of calcifying tissues. Calcif. Tissue Int. 36, 317–319.

PART
3

Global Aspects

17

Biogeochemical Cycles—Minerals and the Origin of Biomineralization

The Earth is thought to be about $4\frac{1}{2}$ thousand million years (By) old. Life evolved on the planet at least $3\frac{1}{2}$ By ago and oxygen began to accumulate in increasing amounts after about 2 By. The first exoskeletons and hard parts did not appear, however, until about 0.6 to 0.5 By ago, so that biomineralization is a relatively new phenomenon, occupying only about the last 10% of the Earth's history (Fig. 17.1). The significance of these events has only become apparent in the past few years as our concepts of the Earth's crust have changed.

I. Early Concepts of the Earth's Crust

In 1715, Edmund Halley, who predicted the activities of the comet named after him, was one of the first scientists to suggest a way to calculate the age of the world. He argued that the age of the oceans and thus of the Earth could be estimated if one knew the salt content of the sea and the rate at which it was added by the rivers. This calculation, of what is now called the ocean residence time for sodium, produces an answer of about 90 million years. Among Halley's other activities was the financing and editing of Isaac Newton's "Principia Mathematica," and the two scientists no doubt discussed these calculations. Newton had a different approach. The traditional view of the world was that it had formed from a hot body of matter condensing in space and slowly

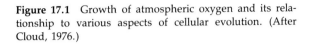

Figure 17.1 Growth of atmospheric oxygen and its relationship to various aspects of cellular evolution. (After Cloud, 1976.)

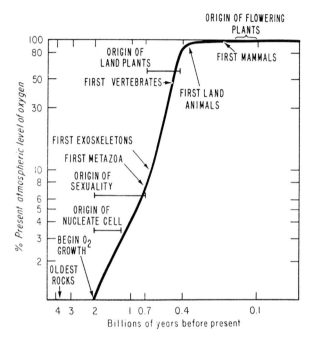

cooling. As it did so, it was envisaged as shrinking, causing the crust to fold into mountain ranges and trenches. Since the cooling rate could be calculated by Newton, he was soon able to show that the Earth would have reached its present temperature in about 100 million years.

It will be apparent that Halley's and Newton's estimates of the age of the Earth differ by a factor of about 50 from current calculations. Why were these estimates so out of keeping with more recent views? The answer is largely related to the discovery of radioactivity. Not only did this allow more accurate aging of the rocks, but it indicated that the decay of uranium, thorium, and potassium in the Earth's core provided a large source of heat that must have slowed its rate of cooling. It also suggested that there may be convective currents of molten rock below the Earth's crust (Wilson, 1971). This discovery that the Earth was a source of heat clearly rendered Newton's calculations inapplicable, but what was wrong with Halley's approach?

Modern calculations of the residence time of various elements in seawater give different times for most elements. Some, such as Al, have a residence time of only 100 years, so it is again clear that such calculations do not measure the age of the sea. Instead they measure the flux of various materials through it.

II. Continental Drift

The material that is weathered from the Earth's lithosphere is carried down rivers and into the oceans where it forms sediments. Calculations show that this material is sufficient to have filled the oceans six times during the Earth's history. Clearly, the world is in a dynamic state and differs considerably from the "rigid Earth" view that Newton had. But how dynamic it is has been one of the scientific surprises of the past 30 years.

   In the 1950s oceanographers established in some detail that there was a continuous range of undersea mountains, usually in the middle of the oceans. Around the same time, Blackett and colleagues (1965) showed that the scattered directions of magnetism of old rocks could be accounted for if it was assumed that the continents had subsequently moved apart. How they did this could be accounted for on the earlier suggestion of Hess (1962) that the sea floor cracks open along the crest of the midocean ridges and that new sea floor forms there and spreads out on either side. On this basis, the continents are envisaged as six major rafts attached to and moving with a spreading sea floor. Where continents meet in areas such as around the Pacific coast of South America the oceanic plate dives under the continental plate to form deep trenches associated with regions of earthquakes. Where the plate turns down, some of the sediment on its surface is scraped off and piled up on the landward side. The rest of the material passes into the mantle, eventually melting at a depth of about 300 km, and finally breaking up and mixing with the molten rocks at a depth of about 700 km to join the crustal material from which it came. Thus, there is a continual formation and destruction of the Earth's crust and this explains why the present oceans contain no sediment older than 150 million years and with very little older than 80 million years.

III. Biogeochemical Cycling

Underwater volcanoes normally begin to arise near spreading centers along which they eventually pass at a rate of from 1 to 10 cm/yr. If a volcano rises fast enough to surmount the original depth of the water and the sinking of the ocean floor, it emerges as an island on which fringing coral reefs are often found. As the system moves slowly toward the subduction zone where one continental plate passes beneath another, submerged volcanoes or guyots are produced and these act as the base for many atolls (Fig. 17.2). Thus, once a base is established in the photic region of a warm sea, a biomineralizing ecosystem becomes established that removes calcium carbonate from the surrounding sea and transports it, with its associated sediments, toward the edge of a plate. This is a clear example of biogeochemical cycling, illustrating the way that biological forces have influenced the state of the mineral world. Other examples can be found in the influence of plants on the flux of water over the Earth's surface, in the breakdown of rocks by algae and lichens, and particularly in the gaseous composition of the atmosphere. A simplified model of some of these influences is given in Fig. 17.3. This shows a region of sulfate reduction, depositing FeS_2 (zone II); a band of methanogenesis based on bacteria using acetic acid as a terminal electron acceptor (zone III); and a layer of nonbiological diagenesis (zone IV). Where these regions are uplifted back into the atmosphere, bacterial oxidations produce strong acids that cause further erosion. An interest-

Figure 17.2 Illustration of crust formation at a mid-oceanic ridge and its subduction where two plates meet (top). The movement of the plate affects the evolution of islands and coral growth.

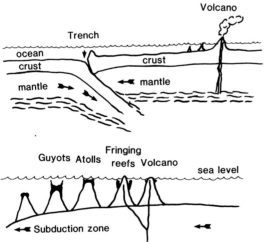

Figure 17.3 Some aspects of biogeochemical cycling. Components of calcium carbonate and silica deposition contribute to the formation of marine sediments (I). Anaerobic bacteria reduce sulfate to sulfide and trap iron (II). Methane-producing bacteria (III) and nonbiological diagenesis lead to the deposition of organic deposits (IV). When these deposits are uplifted into the atmosphere there is erosion and recycling of these deposits. (Based on Westbroek, 1983).

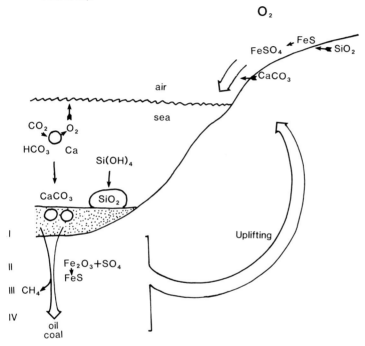

ing variant of these processes has recently been discovered by the U.S. Geological Survey, which has isolated a bacterium (GS-15) that oxidizes acetate to carbon dioxide by converting 8 mol of ferric ion to the ferrous state. These bacteria are anaerobes which produce Fe_3O_4 (magnetite) as an extracellular deposit that probably anchors them in their anaerobic strata (Frankel, 1987). On a biomass basis they are about 5000 times as effective as magnetotactic bacteria, and it is therefore thought that organisms similar to these were responsible for producing the major banded iron deposits that formed on the Earth's surface during the Precambrian (Lovley et al., 1987).

Thus, plate tectonics, continental drift, and biogeochemical cycling are now accepted as the explanation for a whole range of phenomena from the distribution of plants and animals to the structure of the Earth's crust, and from geomagnetic abnormalities to the distribution of corals, fossils, and sediments along the sea bed.

At about the time that these ideas of the Earth's crust were being developed, a number of surveys were being undertaken to monitor the distribution of artificial radionuclides released into the environment by atmospheric nuclear weapons testing. These radioisotopes were of considerable interest, largely because of their possible health effects but also because they represented a unique experimental situation in that they had a single site of entry—the air/ocean interface. Surprisingly, it turned out that the water column tended to be rather deficient in these materials, and short-lived fission products such as ^{141,144}Ce appeared in unusual places such as sea cucumbers 2800 m deep in the sea (Osterberg *et*

Figure 17.4 Settling rates of materials of various sizes in the marine environment. (Redrawn after Degens *et al.*, 1985.)

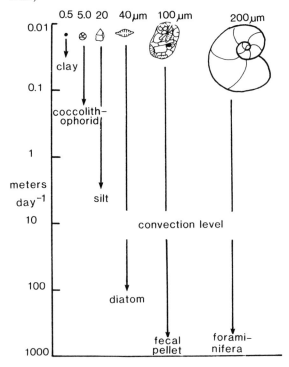

al., 1963). The sinking rates of fallout particles are too low to account for this accumulation and it was suggested that some form of biological incorporation must be involved. Interest centered on various types of inorganic debris.

Subsequent scientific cruises such as the international GEOSECS (Geochemical Ocean Studies) expedition to the Atlantic and Pacific oceans paid particular attention to determining the distribution in surface and deep water of materials associated with particles in the sea. There are four main sources of particles in the oceans: suspended river sediments, dust, cosmic material, and biological materials. By analyzing the material filtered out of thousands of liters of surface and deep water, it became clear that biological particles dominated these contributions (Fig. 17.4). In surface waters organic matter comprised 30 to 70% of the total particulate matter. The hard skeletal phases of marine organisms consisting of $CaCO_3$ and SiO_2 comprised about 25 to 50% of the solid matter. As these particles fell through the water column, a large number of trace elements became adsorbed onto their surfaces while those recovered from sediment traps in the depths of the oceans were corroded and partly dissolved (Lal, 1977). Clearly, biomineralization was an important component for both the downward flux of a whole range of biologically important ions and the formation of sediments that would be recycled into the Earth's mantle.

IV. Oceanographic Cycling

The oceans contain about 10^{13} kg of suspended particulate matter. In fact, for every kilogram of organic carbon produced by the phytoplankton of the sea about 3.8 kg of $CaCO_3$ and 4.2 kg of silica are formed (Fig 17.3). There is, therefore, a massive and continual production of mineral in the oceans, and the rate at which such particles settle through liquids was investigated originally by Stokes, who found that it depended on the differences in density, the viscosity of the liquid, and the cross sectional area of the particles. Because of convection in the euphotic zone of the sea, clay particles (0.5 μm), coccoliths (5 μm), and silt (20 μm) would probably never reach the sea bed were it not for the browsing of zooplankton and the production of fecal pellets (100 μm). These particles, the larger diatoms, and foraminifera form most of the detritus which reaches the sea bed (Fig. 17.4). Of these materials, fecal pellets and plankton hard parts of at least 100 μm in size are now considered to be the most important components of these deposits

in all phyla but it was rapid and dramatic (Fig. 17.7). Furthermore, phosphatic hard parts were more common in the earlier forms than were calcareous ones. The phenomenon has tormented geologists for many years and a variety of suggestions have been proposed to explain it (Simkiss, 1989).

Figure 17.7 Distribution of calcium mineralized tests at the Phanerozoic boundary. Bioinorganic materials: C, calcareous; P, phosphatic. (From Lowenstam and Margulis, 1980.)

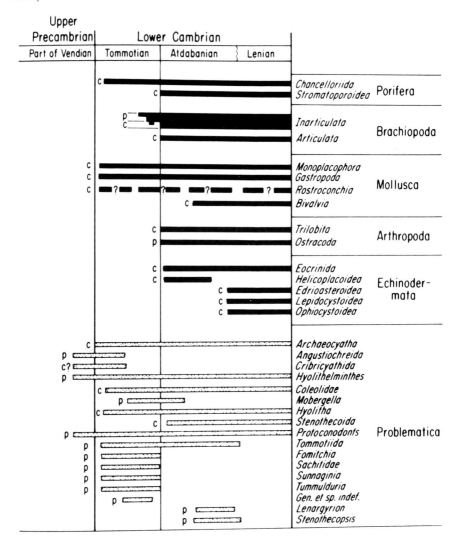

V. Origins of Biomineralization

According to Degens *et al.* (1985), the Precambrian sea contained mainly sodium bicarbonate with little calcium, but this became progressively titrated with volatile acids until the pH dropped to about 8 and a sodium chloride ocean was formed (Fig. 17.8). As the pH of the oceans fell, the calcium level of the seawater started to rise. Initially, noncalcifying organisms were thought to have secreted calcium-binding molecules which would flocculate clay particles and facilitate the ocean cycling of elements. Subsequently, biomineralization developed, largely as a detoxification process that enabled the organism to remove the calcium stress that the influx of this metal was causing.

A second group of hypotheses suggests that changes in the atmosphere were the critical causative agent for the onset of biomineralization. According to Towe (1970), shell and collagen formation are oxygen-dependent processes that were constrained until the level of oxygen in the air rose with its accumulation from photosynthesis (Fig. 17.1). The alternative, namely that it was a fall in carbon dioxide levels, leading to the precipitation of dolomite and a change in the Mg/Ca ratio of seawater is preferred by Riding (1982) as an explanation for the ease of cyanophyte calcification at the end of the Precambrian period.

Perhaps the most plausible of the explanations relates the onset of calcification to the general phosphate metabolism of cells. Phosphate is crucial for nucleic acid metabolism and energy-trapping reactions and normally occurs at concentrations of 0.5 to 20 mM within cells. This is, however, at least a thousand times greater than could exist at the cal-

Figure 17.8 Evolution of ocean chemistry during geological time. (Redrawn from Kempe and Degens, 1985.)

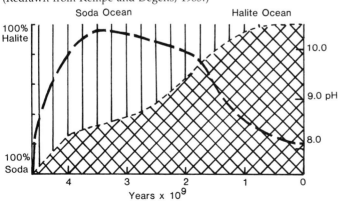

cium levels of 10 mM which occur in many body fluids and marine habitats. To avoid precipitation either calcium or phosphate levels must be reduced in the cell, and since phosphate is essential for so many biochemical reactions it was calcium that was extruded from the cell. The maintenance of a low intracellular calcium ion concentration made possible the evolution of a wide range of calcium binding proteins associated with the transduction of chemical messages (Kretsinger, 1983). These proteins are well conserved, and Lowenstam and Margulis (1980) take this as evidence that these proteins and thus calcium extrusion from cells occurred at least 150 million years before the evolution of skeletalized hard parts. This, however, simply relates biomineralization to the presence of membrane calcium pumps which, as we saw in Chapter 5, is one of the basic requirements of most theories of calcification. It still leaves largely unexplained the sudden onset of this phenomenon 600 million years ago.

An alternative suggestion to explain the onset of large-scale biomineralization in the Cambrian period has been provided by Stanley (1976). He argued that during the evolution of the eukaryotic cell in the Precambrian there was little competition and thus a relatively slow rate of evolution. Once the browsing and carnivorous habitat evolved, however, there would be strong selection for organisms to protect themselves by a variety of skeletal structures, which rapidly appeared in a whole range of phyla. This explanation is a somewhat simplistic one, but it does emphasize how the destruction of hard substrates by browsing and burrowing has probably evolved in association with the exploitation of biomineralization as a way of strengthening jaws and skeletons.

Bioerosion has generally been greatly underestimated by biologists (Warme, 1975). Among microorganisms it is possible to recognize epiliths (which remain on the surface), chasmolites (which penetrates into fissures), and endoliths (which bore directly into the hard substrates). They occur on many sites including limestones, dolomites, corals, molluscan shells, echinoderm skeletons, bones, and teeth (Perkins and Tsentis, 1976) (Fig. 17.9). Among multicellular organisms the effects of eroding organisms are extensive but hard to quantify since such calculations involve estimating the numbers of animals, their feeding rates, and the period of eroding activity (Scoffin et al., 1980). An example has, however, been provided for a fringing reef of 10,800 m² in Barbados. The total productivity of this reef was approximately 206×10^6 g/yr but sponges, parrotfish, and sea urchins alone destroyed more than half of this (Table 17.1). The situation may, however, be even more complex

Figure 17.9 (a) SEM of plastic cast of the green alga *Phaeophila engleri* boring, showing associated sporangial structures. Note imprint of the cross lamellar shell on borewall cast. Gastropod substrate was exposed for 5 months anchored at 5 m depth on the shelf at St. Croix. Scale equals 50 μm. (b) SEM of an endolithic assemblage found within an etched conch substrate after an exposure period of only 9 days. This was the highest degree of infestation detected within such a short period of exposure. The flat surface in the background (arrowheads) represents a plastic cast of the exposed surface of the substrate. The shell was totally dissolved to allow examination of initial endolithic infestation just below the surface. Scale equals 50 μm (Perkins and Tsentis, 1976). (Originally published by the Geological Society of America.)

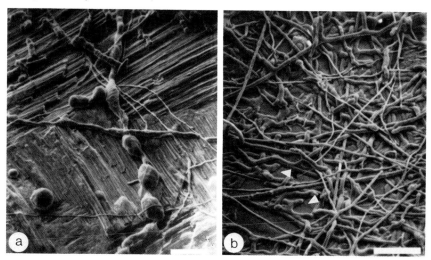

since there is evidence that the initial agents in the erosion of the Great Barrier Reef may be polychaetes, which modify the substrate and make it suitable for other destructive organisms (Hutchings and Bamber, 1985). It appears, therefore, that although living systems are involved in major aspects of mineral deposition, they are also involved in erosion and substrate destruction. Thus, biomineralization plays a major part in

Table 17.1 Erosion and Productivity of a Fringing Coral Reef[a]

Erosion		
Sponges	24.8 ±	4.4
Parrotfish	0.4 ±	0.05
Sea urchins	97.0 ±	5.7
Total	123 ±	7
Gross productivity	206 ±	10

[a] From Scoffin *et al.* (1980). All values are $\times 10^6$ g/yr.

biogeochemical cycling and in the process may have influenced not only the evolution of life but also the form of the planet as we now know it.

References

Blackett, P. M. S., Bullard, E., and Runcorn, S. K. (1965). Continental drift. *Philos. Trans. R. Soc. London, Ser. A.* **258**, 1–322.

Cherry, R. D., Higgo, J. J. W., and Fowler, S. W. (1978). Zooplankton faecal pellets and element residence times in the ocean. *Nature (London)* **274**, 246–248.

Cloud, J. (1976). Beginnings of biosphere evolution and their biogeochemical consequences. *Paleobiology,* **2**, 351–387.

Degens, E. T., Kazmierczak, J., and Ittekkot (1985). Cellular response to Ca^{2+} stress and its geological implications. *Palaeontologica* **30**, 115–135.

Fowler, S. W., and Knaver, G. A. (1986). Role of large particles in the transport of elements and organic compounds through the oceanic water column. *Prog. Oceanogr.* **16**, 147–194.

Frankel, R. B. (1987). Anaerobes pumping iron. *Nature (London)* **330**, 208.

Hess, H. H. (1962). History of ocean basins. *In* "Petrological Studies: A Volume in Honor of A. F. Buddington" (A. E. J. Engel, H. L. James, and B. F. Leonard, eds.), pp. 599–620. Geological Society of America, Boulder, Colorado.

Hutchings, P. A., and Bamber, L. (1985). Variability of bioerosion rates at Lizard Island, Great Barrier Reef: Preliminary attempts to explain these rates and their significance. *Proc. Int. Coral Reef Congr., 5th* **5**, 333–338.

Kempe, S., and Degens, E. T. (1985). An early soda sea? *Chemical Geology,* **53**, 95–108.

Kretsinger, R. H. (1983). A comparison of the roles of calcium in biomineralization and in cytosolic signalling. *In* "Biomineralization and Biological Metal Accumulation" (P. Westbroek and E. W. de Jong, ed., pp. 123–131. Riedel, Dordrecht, The Netherlands.

Lal, D. (1977). The oceanic microcosm of particles. *Science* **198**, 997–1009.

Lovley, D. R., Stolz, J. F., Nord, G. L., and Phillips, E. J. P. (1987). Anaerobic production of magnetite by a dissimilatory iron-reducing microorganism. *Nature (London)* **330**, 252–254.

Lowenstam, H. A., and Margulis, L. (1980). Evolutionary prerequisites for early Phanerozoic calcareous skeletons. *Biosystems* **12**, 27–41.

Osterberg, C., Carey, A. G., and Curl, H. (1963). Acceleration of sinking rates of radionuclides in the ocean. *Nature (London)* **200**, 1276–1277.

Perkins, R. D., and Tsentis, C. I. (1976). Microbial infestation of carbonate substrates planted on the St. Croix shelf, West Indies. *Geol. Soc. Am. Bull.* **87**, 1615–1628.

Riding, R. (1982). Cyanophyte calcification and changes in ocean chemistry. *Nature (London)* **299**, 814–815.

Scoffin, T. P., Stearn, C. W., Boucher, D., Frydl, P., Hawkins, C. M., Hunter, I. G., and MacGeachy, J. K. (1980). Calcium carbonate budget of a fringing reef on the west coast of Barbados. Part II—Erosion, sediments and internal structure. *Bull. Mar. Sci.* **30**, 475–508.

Simkiss, K. (1989). Biomineralization in the context of geological time. *Phil. Trans. Roy. Soc. Edinburgh* (in press).

Stanley. S. M. (1976). Fossil data and the Precambrian–Cambrian evolutionary transition. *Am. J. Sci.* **276**, 56–76.

Towe, K. M. (1970). Oxygen–collagen priority and the early metazoan fossil record. *Proc. Natl. Acad. Sci. U.S.A.* **65,** 781–788.

Warme, J. E. (1975). Borings as trace fossils, and the process of marine bioerosion. In "The Study of Trace Fossils" (R. W. Fray, ed.). pp. 181–229. Springer-Verlag, New York.

Westbroek, P. (1983). Life as a geologic force: New opportunities for paleontology. *Paleobiology* **9,** 91–96.

Whitfield, M. (1981). The world ocean: Mechanism or machination? *Interdiscip. Sci. Rev.* **6,** 12–35.

Wilson, J. T. (ed.) (1971). Continents adrift. *Sci. Am.* 1–172. Freeman & Co., San Francisco, 172 pp.

18 | Overview and Perspective

Biomineralization is an interdisciplinary study. It clearly embraces the chemical laws involved in the formation of a solid phase from soluble precursors. It is induced by direct cellular activity which perturbs the environment either within the organism or in the adjacent ecosystem. The products of these biological activities provide one of the most conspicuous records of life on this planet and the way it has changed over many hundreds of million years. The phenomenon has, therefore, attracted ever-increasing interest from chemists, biologists, and geologists.

I. The Chemical Approach

For chemists, biomineralization poses challenges to the understanding of the way that living systems control surface chemistry. They bring to the problem a wealth of understanding of the principles involved in the induction and growth of mineral solids from pure solutions. If these principles could be directly applied to the very complex conditions that exist in biological systems, then many of the systems that are exploited by cells for the deposition of minerals and the formation of skeletons could be identified. As Mann (1983) writes, "modification of the activation energy barriers for mineral growth and phase-transformation by biological chemical control can, in principle, result in the selection of specific crystalline intermediates . . . ." But, "the most difficult problem in applying this approach to biological systems is that thermodynamics depends upon defining the driving forces that are present. If a biological system can release energy at a small localized site, then that site becomes transiently different from the thermodynamic influence in the bulk solution" (Nancollas, 1983). It is this heterogeneity of the biological system that poses so many problems.

The chemist has clearly identified that one of the essential problems of biomineralization is the control of phase changes. There is a succession of energy steps involved in taking a hydrated ion from solution and producing an organized mineral solid, and to a large extent the phenomenon has to do with dehydrating ions and forming new surfaces. Biological systems actually seem to be remarkably good at both these activities. It is quite likely that there may be several different ways to form and control the growth of a mineral surface; but it is now a defined experimental problem and there is no doubt that chemists will make major contributions to this aspect of biomineralization.

II. The Biological Approach

At the cellular level, biologists have also identified a number of the essential requirements for the deposition of minerals by living systems. Biomineralization usually occurs in a small enclosed space within which it is possible to surpass the solubility product of a mineral by changing the concentration of precursor ions. The small volume enables the solubility product to be exceeded with a minimum involvement of the energy-dependent processes of active transport or anion synthesis. It also reduces the "buffering" effects of large volumes so that the process can be easily regulated. Frequently, the site of mineral deposition occurs within an organism but in some forms materials move out of cells to bring about mineral formation on an outer cell or epithelial surface.

The products of biomineralization are often shaped in elaborate ways and are atypical of the crystals produced by purely inorganic processes. We need to know how this cellular control is exercised to produce deposits that are commonly of very precise size and uniform crystal orientation. The diffusion of ions through solutions and over crystal surfaces tends to be a random process whereas biological shapes are clearly the products of nonrandom activities. At what level is the biological control of shape introduced into the system? Is it by regulating at the molecular level the sites at which ions are introduced into the mineralizing structure? Is it by using the properties of polyelectrolytes within the mineralizing fluid? Or is it by cellular control of the shape of the space in which mineralization occurs? These possibilities have now been clearly identified but await experimental analysis.

The contribution that biologists have made to the study of biomineralization is twofold. First, they recognize that cells only survive because they can regulate the influx and efflux of ions, either by influencing the passive permeability of cell membranes or by actively transporting ions by energy-dependent pumps. The driving forces for biomineralization are part of this basic phenomenon of biology. Second, biologists are acutely aware of the enormous and specific information content that can exist in protein macromolecules, and they are very familiar with the concept of self-assembly. The transfer of information by surface interactions underlies many of the phenomena of cell biology and these ideas are clearly relevant to biomineralization. Thus, the two concepts of "pumps versus proteins" underlie virtually all biological discussions of biomineralization and, as simplistic generalizations, they are extremely helpful. One of the major purposes of this book, however, has been to indicate the diversity of the biological processes that meet the general requirements for converting ions into mineral deposits. And there is no doubt that they are very diverse indeed. As the recognition of this diversity of minerals and cellular systems increases, the biologist will need to realise two things. First, there is an enormous wealth of experimental material available for study of any particular aspect. Second, it is unlikely that only one fundamental process is involved in all these systems. The study of biomineralization has probably reached the stage where it would be advantageous for a while if biologists were to study fewer systems in greater detail in order to characterize their properties more precisely.

III. The Geological Approach

Geologists bring to the study of biomineralization an extended view of biological time, posing questions about extinct forms, the succession of minerals found in fossil organisms, the relationships between biomineralization and the ancient environment, and what aspects of biomineralization have led to its selection over prolonged evolutionary time scales. In discussing the origins of biomineralization, Lowenstam and Margulis (1980) comment, "there is no reason to believe that active genetically determined intracellular calcification occurs in prokaryotes."

There is, however, evidence that bacteria have been able to form magne-tosomes since before the Cambrian. As Kirschvink *et al.* (1985) deduce "from the wide spectrum of organisms with the ability to precipitate magnetite," it "probably evolved prior to the major radiation of animal phyla in the late Precambrian." Should one therefore consider magnetite as the primordial biomineralization product? Is it controlled by a specific gene and was there subsequently a gene for intracellular calcification?

Most of the speculations about the origins of calcification relate it to the low intracellular concentration of calcium ions. According to Degens *et al.* (1985), biomineralization was a response to cellular overloading of calcium in the Precambrian oceans, while Lowenstam and Margulis (1980) see it as a way of protecting the calcium-sensitive proteins of the cell. The central concept in both of these schemes is a transmembrane pump based on a Ca-ATPase. Such calcium pumps have been exten-sively studied in recent years, although relatively little attention has been given to their relation to mineralization. This is unfortunate since it is now clear that there is a great variety of such molecules. We would strongly advocate further study of the Ca-ATPase systems of the plasma membrane and intracellular membranes, if possible with intact organ-elles, as a method of studying biomineralization. The relationship of cell membranes to intra- and extracellular mineralization is a crucial area of study and clearly relates to the mechanisms of calcification.

Geological speculations on the origins of biomineralization can, there-fore, be a strong stimulus for further experimentation. In recent years geologists have made two other major contributions. The first is the recognition of the great diversity of inorganic products that living sys-tems produce. The ever-lengthening lists of "biological minerals" which Lowenstam has compiled have extended the interest in biomineraliza-tion from a study of silica and calcareous products to one that includes at least 15 elements and over 60 mineral forms. Inevitably, this has resulted in a change in perspective. We are clearly no longer considering just a few cellular pathways involved in skeletogenesis but are now concerned with much of the subject and diversity of inorganic biochemistry. Redox potentials were relatively unimportant in traditional studies on the for-mation of calcareous or siliceous skeletons but they are clearly dominant influences in the formation of manganese and iron deposits. The realiza-tion of the range of elements involved in biomineralization processes also underlies the second major contribution of geologists. Living pro-cesses can now be seen as having had a major influence on the composi-tion of Earth strata, and on the biogeochemical cycles that are involved in much of the geological activity on the Earth's crust. The intellectual and commercial applications of these concepts are still being assimilated by both academic and industrial scientists.

IV. The Components of Biomineralization

It will be apparent that, although a number of different disciplines have contributed to the study of biomineralization, there is general agreement about the fundamental processes involved. This is particularly true at the biochemical level but probably less clear at the cellular level.

A. Biochemical Aspects

The general concept of biomineralization is that it is induced by pumps and controlled by matrix effects. Pumps, or more precisely the enzymatic increase in the activity of ions at sites of mineral deposition, have been difficult to demonstrate *in vivo*. There are, however, a range of inhibitors which affect ion-motive ATPases and which could be used to distinguish between these diverse enzymes (Pedersen and Carafoli, 1987). Certainly, there is a critical need to establish which ions are actively moved in particular forms of biomineralization. There is every reason to believe that the pumps involved on plasma membranes and vacuolar membranes differ, although both systems may be involved in biomineralization. A close association between basic cellular activities and mineral deposition is one of the reasons for believing that the phenomenon results from an association of a few fundamental properties of cell biology. It is also one of the reasons why it is difficult to study the process without disrupting normal biochemical pathways. The three most common ion pumps (H^+-ATPase; Na^+/K^+-ATPase; Ca-ATPase) are all involved in fundamental cellular processes such as energy transduction, osmoregulation, and intracellular signaling. Any one of these enzymes could, however, be used to trigger biomineralization, e.g.,

$$Ca^{2+} + 2HCO_3^- \xrightarrow{H^+\text{-ATPase}} CaCO_3$$

could trigger mineral deposition by removing protons from the site of calcification; or

$$Ca^{2+} + CO_3^{2-} \xrightarrow{Na/K\text{-ATPase}} CaCO_3$$

could induce mineral formation by solute-linked water movements. The movement of these ions can cause an osmotic flow of water, increasing the concentrations of the reactive ions left behind at the mineralizing site; or

$$Ca^{2+} + CO_3^{2-} \xrightarrow{\text{Ca-ATPase}} CaCO_3$$

could initiate calcification by increasing the concentration of calcium ions at a localized site. The problem is, therefore, to find ways of studying these processes experimentally.

The study of the role of the organic matrix in biomineralization has been equally clear but difficult to study. The most popular current view is that an insoluble matrix forms a structural framework to which a soluble matrix attaches and acts as a nucleating surface. If the soluble matrix exists in solution, it may have a different effect and act as a mineralization inhibitor. This can occur either by binding ions in solution so as to reduce their charge and mobility or by attaching to crystal nuclei and surfaces, blocking the possibility of further crystal growth. If attached to a solid phase, the same matrix may act as a nucleating surface facilitating mineralization. These are both ingenious concepts and difficult systems to manipulate experimentally. Yet it is crucial to many schemes of biomineralization that we obtain good data on the composition and properties of these organic surfaces and one of the features that needs most consideration is the relationship of the structural (insoluble) matrix and the reactive (soluble) matrix.

The role of water in the biomineralization process is also in need of clarification. Most of the effects of membrane pumps in increasing ion concentrations could, as we have already pointed out, also be achieved by reducing the concentration of water present. The presence of water, however, influences many other aspects of the biomineralization process. In biological systems, most ions are heavily hydrated and often have to be partially dehydrated in order to traverse the cell membrane. They are also often dehydrated when incorporated into mineral crystals and there is a considerable energy involvement in achieving this state and in forming a solid surface. Water is certainly not just a passive solvent and more attention should be paid to its reactive role in biomineralization (Taylor & Simkiss, 1984).

B. Cellular Aspects

It is possible to define biomineralization in its broadest sense as being the formation of minerals by biological influences. These influences will vary according to how the biological structures are organized, and

Wilbur and Simkiss (1979) recognized three systems, namely, individual cells, epithelia, and syncytia. An elaboration of these systems is given in Table 18.1 and it has two attractive features. First, it emphasizes the relationship between cells and mineral deposition and, second, it implies that as cellular systems become more elaborate they increase the possible ways of forming inorganic deposits. This latter point is illustrated in Fig. 18.1. With a single cell, the various biochemical events that induce mineral deposition (calcium pumps, carbon dioxide removal, H^+/OH^- changes, etc.) could be oriented outward so as to have an environmental interaction, or inward so as to induce intracellular mineralization (Fig. 18.1a,b). Putting several cells together to form an epithelium produces a number of additional systems (Fig. 18.1d,e). First, the activity of the cells and pumps could be oriented in various directions producing, for example, an apical system of mineral deposition and a basal system of proton excretion or calcium uptake. Second, there are the possibilities of intercellular ion movements as well as intracellular ones for both secretion and resorption. These will, of course, have quite different requirements in terms of cellular energetics since the intracellular and extracellular routes have quite different ionic components. Third, since the multicellular organism contains its own blood or "milieu intérieur" it is quite likely that some cells quite distant from the site of mineral deposition raise the ionic concentrations of the body fluids to levels corresponding to supersaturation levels for various minerals. In that situation, the driving force for extracellular biomineralization may be quite distant from the site of mineral deposition. Moreover, the whole

Table 18.1 Types of Mineralizing Systems in Various Taxa

| | Single Cell | | Syncytial | Epithelial | Glandular |
	Extracellular	Intracellular			
Bacteria	+	+			
Algae		+			
Higher plants	+	+			
Protozoa	+	+			
Porifera	+	+			
Bryozoa				+	
Coelenterates		+		+	
Annelids					+
Crustacea		+		+	
Mollusca		+		+	
Echinoderms			+		
Chordates	+			+	

Figure 18.1 Types of cellular systems involved in biomineralization. A generalized cell is shown on the left with systems for transporting calcium ions and protons and fixing carbon dioxide. These cellular activities could interact with the surrounding environment to induce extracellular mineralization (a). Alternatively, if ions are transported into a vesicle they may produce intracellular mineralization (b) or part of the cytoplasm may be shed to produce a mineral-containing matrix vesicle (c). The grouping of cells together produces a "milieu intérieur" with a regulated ionic composition from which mineral may be deposited by an epithelial system using either intercellular or intracellular routes (d). Finally, the fusing together of cells to form a syncytium produces a collaborative cellular system of biomineralization (e). In all cases, mineral deposits are shown in solid black, carbon dioxide is shown being fixed by chloroplasts, and the processes of biomineralization are illustrated by calcium. Similar concepts could, however, be applied to most forms of biomineralization.

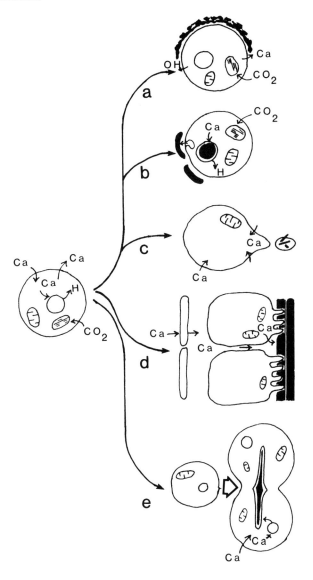

process could be regulated by the presence or absence of various inhibitors. Finally, it is worth emphasizing that epithelial systems of biomineralization not only imply a diversity of routes and activities for the production of the mineral but also a variety of physiological differences in the epithelial cells. It is an oversimplification to consider the epithelium as an entity. Rather, there are physiological differences between the cells of those epithelia and this should become an interesting focus for future work. This aspect of cellular interaction is, of course, at its extreme with syncytial systems of mineralization (Fig. 18.1e). It is seen in a number of ways. The formation of vesicles, which transport intracellular material to a site of mineralization, the "communal" region where crystal deposition occurs, and the way that cells enter and leave this syncytial site are clearly unusual aspects of cell biology. The cooperation between individual cells is extended in forming spicules such as those of the larval sea urchin and is all the more remarkable because these cells produce such well-organized crystal structures. It is difficult to believe that there is not feedback from the crystal surface to the cell that is forming it in order to maintain this level of organization. This aspect of cellular organization is, however, completely unexplored and, in order to emphasize the fundamental contribution of cell biology to this process, we have included the concept of cellular pathology in Fig. 18.1c. The possibility that cells can respond to an excess of intracellular calcium by shedding a calcium-rich part of the cytoplasm not only provides an interesting speculation on the origin of vertebrate matrix vesicles but also reminds us of how little is understood about bone formation and what an unusual form of biomineralization it represents. Clearly, the organization of cells and their basic properties remain a largely unexplored aspect of biomineralization.

V. Future Studies on Biomineralization

Biomineralization is largely about ions and structures, initially at the molecular and cellular level and subsequently at the organismal level. As such, it should be at the forefront of studies on molecular biology and biotechnology. The truth is, however, that it has suffered from an absence of a clearly defined set of testable hypotheses and an inconsistency in support. This is likely to change in the next two decades for two reasons. First, the general advance in the interpretation of biological phenomena at the molecular level has already produced a whole range of new methods which, simply by their application, will throw new light

on biomineralization. Second, the use of biological processes in the production of industrially important minerals is almost certain to increase.
The attraction of growing crystals in zero gravity or of purifying mineral
products under carefully controlled conditions will cease to exist if it is
possible to prepare a cell culture that can achieve the same result at
minimal cost.

It can be said with considerable confidence that our understanding of
the mechanisms of biomineralization will be aided and expedited by
application of the techniques of molecular biology. By means of recombinant DNA technology, information on the regulation of the synthesis of
proteins having a specific function can be obtained more easily and with
less ambiguity than by working directly with proteins (Prockop, 1981;
Ludwig and Volcani, 1986). This approach should be especially valuable
in investigating proteins that are thought to initiate crystal nucleation, to
control crystal orientation, and to limit crystal size through inhibition.
The methods as they apply to diatoms and silicification have been indicated by Ludwig and Volcani (1986) and include the construction of
complementary DNA (cDNA) "libraries" from mRNA for use to identify
specific genes in total gene banks and then to purify them. The property
encoded by a specific gene can then be determined by showing that the
function is absent when the gene is mutated and returns on replacing
the normal gene. This approach has already been applied to the exciting
topic of cell lineages by using the micromere-mesenchyme cells of the
sea urchin embryo that produce the initial calcareous spicule of the
skeleton. A strand of DNA has been isolated which codes for the matrix
protein of the spicule. This DNA appears to be transcribed into RNA a
few hours before spicule formation starts. The effect increases over 100-
fold during development and is specific to this cell line. This gene apparently contains a single intron and produces a protein with a proline-rich
domain, a very basic C-terminal region, and a tandemly repetitive domain of 13 amino acids which comprises 45% of the length of the protein
(Benson et al., 1987). The presence of glutamic and aspartic acids with
the absence of any basic amino acids suggests this sequence could be
involved in binding calcium ions.

The ability to identify, translocate, or annihilate specific genes therefore provides the necessary techniques for perturbing a biological system. The effects that are produced can then be studied by a host of new
physicochemical measurements. Proton microprobe analysis, nuclear
magnetic resonance, neutron diffraction, small-angle scattering of X-ray
sources and high-resolution electron microscopy are already available
and are being applied to these problems. This increased resolution at the

molecular level will enable us to obtain more accurate data on the organic/inorganic interaction. There is, for example, every indication that the concept of epitaxy in biomineralization has been too rigid a view of this association. Because of the surface energy conditions at the site of nucleation, crystal nuclei often have a slightly different composition from the main crystal phase. As a result, the steric requirements necessary for the matrix to induce mineral formation may be slightly different from those of crystal growth. Some ability for movement between the structural and reactive components of the organic matrix may therefore be desirable, and it is interesting to note that modulators (such as the protein osteonectin) attached to the structural proteins could serve such a function. Alternatively, it is possible that the matrix may bind ion clusters or nuclei rather than individual ions (Termine *et al.*, 1981; Wheeler et al., 1987). As the analytical resolution of the techniques that are available increases, we should be able to get direct data on these suggestions. A similar increase in the ability to study the short-range structure of amorphous minerals should also make it possible to study these materials at the atomic level. It is important to realize that at least 25% of the products of biomineralization are amorphous and these deposits have different physical and biological properties from the more commonly studied crystals.

Finally, there is a whole range of biological aspects of biomineralization that are, as yet, almost unexplored. In most, if not all, phyla mineral deposition proceeds in increments rather than as a continuous process. The evidence is clearly seen as rhythmically formed marks in the skeletons of corals, molluscs, brachiopods, barnacles, and fish otoliths. These changes in the ratio of mineral-to-matrix composition may reflect changes in endocrine activity, the arrangement of the crystal tablets, the effects of dissolution, or various environmental influences. The phenomenon is used extensively by scientists as diverse as geologists and population biologists who interpret these bands as annual markers or growth indicators. The physiological basis of these increments is, however, not well understood. In fact, the whole concept of the control of mineral growth as a skeletal phenomenon is not at all clear. Those invertebrates with a rigid exoskeleton, such as many of the crustacea, can be committed to a system of molting with a total replacement of the skeletal system at the time of growth. Most animals, however, maintain a permanent skeleton with plates or valves that can be extended incrementally, or tubes that can be added to at their ends (Table 18.2). In addition to these variations within an organism, there are numerous interactions between organisms as in the effect of symbiotic associations upon min-

Table 18.2 Examples of various growth processes in invertebrate skeletons

Type of skeleton	Mineral deposition sites	Taxa
Two valves	Edges of valves/increasing interior volume	Bivalve molluscs Brachiopods Ostracods
External plates	Edges of plates/increasing interior volume	Echinoderms Barnacles
Mineral tube	Lips of tube or shell, increasing length and volume	Annelids Gastropods Nautiloids

eral deposition. This occurs frequently between photosynthesis and calcification or, even more surprisingly, between photosynthesis and silicification. These are clearly problems of fundamental interest.

It is to be hoped, therefore, that studies on biomineralization in the future will address some of these basic biological questions. The phenomenon of biomineralization may be rooted in molecular biology but its effects in terms of niche production, ecosystem stability, and geochemical cycling are clearly fundamental to the survival of whole environments.

References

Benson, S., Sucov, H. Davidson, E., and Wilt, F. (1987). A lineage specific gene encoding a major matrix protein of the sea urchin embryo spicule. *Develop. Biol.,* **120,** 499–506.

Degens, E. T., Kasmierczak, J., and Ittekkot, U. (1985). Cellular responses to Ca^{2+} stress and its geological implications. *Palaeontologica* **30,** 115–135.

Kirschvink, J. L., Jones, D. S., and MacFadden, B. J. (eds.) (1985). "Magnetite Biomineralization and Magnetoreception in Organisms," pp. 1–682. Plenum Press, New York.

Lowenstam, H., and Margulis, L. (1980). Evolutionary prerequisites for early Phanerozoic calcareous skeletons. *Biosystems* **12,** 27–41.

Ludwig, J. R., and Volcani, B. E. (1986). A molecular biology approach to understanding silicon metabolism in diatoms. *Syst. Assoc. Spec. Vol.* **30,** 315–325.

Mann, S. (1983). Mineralization in biological systems. *Struct. Bonding, (Berlin)* **54,** 125–174.

Nancollas, G. H. (ed.) (1982). "Biological Mineralization and Demineralization," pp. 1–417. Springer-Verlag, Berlin.

Pedersen, P. L., and Carafoli, E. (1987). Ion motive ATPases. 1. Ubiquity, properties and significance to cell function. *Trends Biochem. Sci.* **12,** 146–150.

Prockop, D. J. (1981). Recombinant DNA and collagen research. Is amino acid sequencing obsolete? Can we study diseases involving collagen by analysis of the genes? *Collagen Res.* **1,** 129–135.

Index